Defining Digital Architecture 数码建筑

刘育东/编

大连理工大学出版社

contents 目 录

Digital Architecture? Digitality of Architecture

Before we begin, let me tell you a story. When the first telephone was invented in human history in 1929, the advanced telecommunication that shortened global communication distance triggered a technology revolution; Scribners Magazine from New York published a message that said:

The average human being of to-day is not impressed by miracles.He reads in a newspaper that plans are being made to connect New York with Tokyo by telephone. I doubt that its practical,: he may remark. But the next day he discovers that the thing has actually been accomplished. The day after that he himself calls up Tokio and, if there happens to be a few minutes delay in putting the call through, he complains bitterly about the service.

Whenever the new technology threatens the market of the existing technology, there will always be lots of oppositions. The criticism which hear the most often is that the quality of the new technology is not steady.

If You Are Ready, You Are Already Too Late   Today the importance of computer is no more debatable. In the early 90's, most of the architects have used CAD for the architectural drafting. At that time, comparing to the 2D drafting, Modeling and 3D animation is a quite new area. Most people are still discussing what the computer can do besides drafting.

刘育东
中国台湾交大人文社会学院副院长
国际数码建筑设计奖评审委员会召集人
Yu-tung Liu
College of Humanities and Social Sciences,
Chiao-Tung University
Vice Dean
Committee of International Digital
Architectural Design Award Convoker

何谓数码建筑？谈建筑的数码性

在本文开始之前，让我先讲一个故事。在1929年人类历史上出现了第一部电话，人类通讯即将因全球人与人沟通距离缩短而引发革命时，纽约的《Scribners》杂志刊登了这样一段话："一般人在今天并不会对奇迹留下深刻印象……如果某人从在报上得知一项用电话连接纽约与东京的计划，他先会说，'我怀疑这项计划是否真能实现'。但如果第二天他发现这个新玩意儿竟然已经被发展完成，第三天他就会亲自打电话到东京试试看，在尝试过程中只要等待接通电话的时间有几分钟延误，他就会不假辞色地严厉批评这项新发明的种种问题。"

就像这样，在人类文明发展过程中，每每有新科技发明而足以造成人们日常生活的重大变革时，只要这个新科技普及到可能引发一场"革命"，并对其他相对旧有技术造成威胁时，许多反对这项新科技的声音就会立即出现。而且，最常听到的批评就是，在新科技的品质上这里有问题那里又有问题。

等我们准备就绪时，我们早已落后 和10年前相比，电脑在建筑设计上的重要性已不再需要辩论。20世纪的最后一个10年，利用电脑绘制各类平面图、立面图、剖面图、细部图等二维空间图集已成为全球事务所的标准方法，而另一方面，电脑模型与动画则尚处起步阶段，在这段时间内，讨论的焦点在于"电脑除了是绘图工具之外，对建筑还有什么影响"。

10年后的今天（21世纪的第一年），电脑除了提供所有过去传统媒体的功能，而且将这些媒体变得十分容易操作以外，还承续了建筑历史上有关设计过程与方法的脚步，发展了许多从未出现的设计媒体，如虚拟实境，自由形体技术、网际网络等等。而且由于这10年内有 Frank Gehry、Peter Eisenman 等成熟而又知名的建筑师的带领，Daniel Libeskind、Tom Mayne、Greg Lynn、Eric Owen Moss 等中生代建筑师的发挥，以及 OCEAN、UN Studio、NOX 等等新生代建筑团体的实验，在21世纪之初，建筑设计

Ten years after, today the application of computer has changed the design method, especially the technology like the virtual reality, free form technology, and internet. Through the computer, many famous architects for instance, Frank Gehry, Peter Eisenman, Paniel Libeskind, Tom Mayne, Greg Lynn, Eric owen Moss have designed the amazing space. Many of the architectural elements have been redefined: function. form. volume and space. The new type of architecture predominated by the architect and computer digitized solution is generally called digital architecture. And the discussion of the computerization has been shifted into the digitality of architecture such as "Is this a temporary phenomenon or a permanent revolution?" Will this tool hurt the architecture since we're not familiar with it?" "Are we ready?" "What's digital architecture?"

Actually I still don't have any convincing answers for these questions. Here I can only quote the sentences written by linguist Susan Herring in her essay about the computer- mediated communication:

Rather than wondering whether CMC scholarship is legitimate, a more appropriate question now is how scholarship can best keep pace with the continued expansion and diversification of CMC.

The relationship between the digital architecture and architecture, is similar to the relationship between CMC and linguistics. Another quotation emerges in my mind is by an oriental intellectual, originally quoted by Peter Eisenman in his speech in Taipei,China: "If you are ready, you are already too late."

由于"电脑"这个"数码媒体"设计思考的加入，引发了比以往更惊人而绝对不能再忽视的建筑发展。我们所熟知的建筑基本要素—功能、形式、体量、空间，均已被重新定义。这些由建筑师使用电脑数码化程序在设计过程中所产生的新现象，在国际上被越来越多人统称为"数码建筑"(digital architecture)。尽管如前面所提及，在任何可能由科技所引发的革命最初期，总有正反两种意见并现，但随着时间演进，电脑科技日新月异。现阶段，在建筑领域中讨论的焦点已由电脑的工具性提升为建筑的数码性（digitality of architecture），特别是像下列这样的问题："这是一场建筑大革命还是新工具所引起的短暂现象？"、"我们尚未完全了解就投入，是不是太危险或者会伤害建筑？"、"我们准备好了吗？"、甚至"什么是数码建筑？"等等。

对于以上问题，事实上我仍没有足以令人信服的答案。但看着电脑与网络发展快速到使建筑业与教育界远远被抛在后面时，我脑海中浮现出两段话可供作答。语言学家Susan Herring在1996年针对"以电脑为媒体的沟通"(computer-mediated communication, CMC)在语言学领域引发论战时提及："我们已无需再争论CMC在语言学中是否是个值得深入研究的课题，目前该问的是，我们的研究与讨论如何才赶得上CMC持续快速的扩展与变化"。

同样的道理，数码建筑在建筑中的定位和CMC在语言学中的定位很类似。我的另外一个回答是，去年底（2000年12月2日）Peter Eisenman在中国台北举办的远东交大建筑讲座中，所引述的东方智者的话："等我们准备就绪时，我们早已落后"。（if you are ready, you are already too late.）

数码建筑的内容　虽然建筑在数码时代的发展仍属初级阶段的初期，在这时候为数码建筑下定义，甚至讨论未来它的内容可包括哪些项目，似乎显得有太多的预测性，但如果以电脑自1963年具备绘图能力，以及自1990年Frank Gehry巴塞罗那鱼

The Contents of Digital Architecture   It seems too early to define the digital architecture since it's still in the early stage. Concluding from the drafting software of 1963 and the Barcelona Fish designed by Frank Gehry, we can still generally define the digital architecture as: "Architecture which involves the application of digital media into any phase of designingconceptual design, design development, detail design, construction documents and processand produce critical result no matter in function, form, volume, space or concept." The following question is, what's the future of digital architecture and how shall we architects position ourselves.

To answer these questions, we have to review the history of how the computer influence the architecture. From 1963, architects started to use computers to draft (computer-aided drafting), while the computer is also used to store and process many design datum (computer- aided design).

Up to 1990, the 3D animation technology makes the computer as design media which directly influences the design process (computer-mediated design or design with computer), like what Frank Gehry and Peter Eisenman have been doing. From 1993, the internet wave brought even greater influence on the architecture. The globalization has formed the internet-aided design and web-based design, while the cyberspace and networked space has overthrown the traditional spatial concept. In 1995, the computer modeling is getting sophisticated, there come more and more free forms, beyond our imagination, forming the virtual architecture and virtual space.

Reviewing these phases, we can generally define the architecture using the digital technology as digital architecture.

(Barcelona fish)的设计中所引发建筑的巨变来看，我们还是可以将"数码建筑"在2001年初的发展初步归纳为："凡是将各类电脑数码媒体，使用在建筑设计的以下过程诸如设计概念、早期设计、设计发展、细部设计、施工计划、营造过程等任何一个阶段或几个阶段甚至全部的过程，并因而在功能、形式、体量、空间或建筑理念上有关键性成果的建筑，均可广义地视为数码建筑"。有了这样一个初步定义，接下来的问题则是：全球数码建筑正在迅猛发展，数码建筑可能的前景是什么？我们该如何定位？

让我先用一些篇幅简述电脑在建筑设计发展中的历史，以便用比较精确的用语试着来回答这些问题，并讨论数码建筑可能包含的内容。1963年电脑开始具备绘图能力，将电脑绘图(computer graphics)的功能应用到建筑设计，便成了建筑中常听到的第一代CAD——电脑辅助绘图(computer-aided drafting)。同时自20世纪60年代起，将电脑在人工智能上的发展应用到建筑设计思考过程中，以及用电脑绘图功能来记录大量"建筑图文资料"（构建了所谓的电脑中的"建筑知识"），便进一步被称为"电脑辅助设计"(computer-aided design)，这一直到今天仍是重要的发展方向。1990年起，电脑的数码影像处理与合成技术日渐成熟，再加上电脑模拟与动画能力发展也趋健全，此时电脑在许多设计学科中不再只是被视为"工具"，而是进一步成为思考与表现设计理念与操作方式的"媒体"(media)，因此出现了电脑媒体化设计(computer-mediated design)，甚至被直呼为电脑设计(design with computer)。Frank Gehry与Peter Eisenman等人的设计革命，便属于这一类。另外，自1993年互联网络浪潮带给建筑业更大的冲击。一方面，将全球的距离完全拉近以及将全球的建筑知识（在网络上随时可得的全球建筑图文资料）完全整合，形成所谓"网络辅助设计"(internet-aided design)或"网际设计"(web-based design)，另一方面，网络所形成的网际空间(cyberspace; networked space)，更颠覆了我们建筑向来以点、线、面等几何关系所构建的空间概念。大约自1995年起，电脑在自由形体(free forms)的塑造与操作上渐趋成熟，再加上能呈现更逼真设计的虚拟实境技术也更成熟，越来越多以前想像不到的建筑或空间可被建构，甚至有些在此基础上建造起来，形成所谓的"虚拟建筑"(virtual architecture)与"虚拟空间"(virtual space)。

The Definition of Digital Architecture: New Tool? New Theory? Or New Revolution?   Now it's clearly understood that the computer has influenced many facets of architecture: drafting, image, modeling, animation, multi-media, internet, free form, virtual reality. It can also be predicted that the influence will be just deeper and further. Based on the previous review. I have predicted the following possible visions for the digital architecture:

Digital architecture might be just a new tool. If the digital technology does not influence the design thinking, design method and spatial theory, then it can be treated as merely a new tool. Even though, the digital technology will be another breakthrough as a tool, much more powerful than the drafting invented in the Greek time or the modeling in the Renaissance.

Digital architecture might be a new theory. If it can be used to assist the design thinking process by internet-aided design and web-based design, the design method based on the Bauhaus pattern will be revoluted. In the same time, the design theory from the Renaissance will be also greatly changed by the computer-aided design and design with computer. In addition, if our cyberspace and networked space experience can ultimately influence our spatial concepts in the physical world, following the massy Egyptian space, geometric Greek space, mystic Gothic space, dynamic Baroque space, modernism space, there will be a brand new spatial theoryso called digitalism. On the other hand, if the design method, thinking pattern and spatial theory are all changed by the digital architecture, digital architecture will be more than a theory.

最后，为了反映数码电子时代(简称数码时代)对人类所引起的巨大变革，自1963年电脑对设计各阶段与各层面的冲击，在建筑上均广义地视为建筑在数码时代各方向的发展，并称这些具有数码性(digitality)的建筑为数码建筑(digital architecture)。

数码建筑的定位：新工具？ 新理论？ 新时代？ 新革命？ 由前述电脑在建筑的发展上看(很难想像电脑与建筑的结合，竟会衍生出这么多专有名词，由此也可想像电脑在建筑中发展与更新的迅速)，数码技术包含了绘图、影像、模型、动画、多媒体整合、网络、自由形体技术与虚拟实境等，可以预见，这些技术在以后将会更快速发展并更全面影响建筑。但若要在此刻为数码建筑定位，并了解它的前景(vision)到底是什么，(这么多人辛苦实验、探讨、突破、甚至花费比一般建筑更高的费用，到底他们看到了未来的什么？)于是我可以基于上述回顾来预测(而且仅只是预测)，数码建筑在人类的建筑发展史上，有下列几种可能性：

数码建筑可能只是一种新工具。如果所有前述已成熟的数码技术，仅代表着"电脑绘图"或"电脑辅助绘图"的功能，不具备任何设计思考过程、设计方法以及空间理论上的意义与后续发展机会，则电脑真的只是新工具。而且，作为工具的数码技术是另一个新的突破，影响力将大大罗马时期的平面式绘图与文艺复兴的建模法。

数码建筑可能是一个新理论。如果目前的数码技术具备了辅助设计思考的能力，而又能利用网络来辅助设计方法与设计过程(即internet-aided design与web-based design)，则建筑设计起自20世纪30年代包豪斯的设计方法理论应会大量改写。另一方面，若电脑可作为重要的设计思考媒体，进而透过这个数码媒体的特质(即computer-aided design与design with computer)，产生前所未有的设计思考模式与在此新模式下的建筑物，则我们将会见到自16世纪文艺复兴早期的设计思考理论，在停滞了500年后有了重大发展。最后，若网际空间的经验可能影响到对实质空间的经验 (即cyberspace与networked

The digital architecture might thus form a new age if it is considered to have an overall impact on architecture. We should pay attention if the digital architecture really forms an age, it will influence not only on architecture, but also the value system (Why it's good) and new aesthetics (what's beautiful).

Digital architecture can be defined as a revolution. Every revolution changes human history and life style: fishing and hunting revolution, agricultural revolution and industrial revolution. Since architecture is only a small part of the social evolution. We can only wait to see what's the drastic change in the digital age.

Besides these possibilities, we can also define the digital architecture by its duration. The duration of a new tool is generally a few months, 10 years to 30 years for a new theory, hundreds to thousands of years for a revolution. If the digitality of architecture can remain a few years, it can be defined as a new tool; more than 10 years, it can be defined as a new theory, more than that, it can be defined as a new age, more than hundred years, it can be defined as revolution.

下图： "数码自然"中国台北公信电子公司大厅设计，交大建筑所刘育东、李元荣、施胜诚、黄士诚、黄国贤、张嘉伦、赖德、范扬铮，2000 年 11 月～2001 年 4 月 / BELOW: "Digital Nature", the reception hall of Bcom Electronic, INC. Yu-Tung Aleppo Liu, Yuan-jung Lee, Sheng-cheng Shih, Shih-cheng Huang, Kuo-shang Huang, Chia-lun Chang, Te Lai, and Yang-chen Fan. November 2000 to April 2001

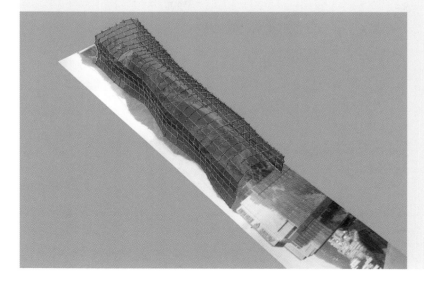

space），而且自由形体与虚拟实境可以营造出以前无法想像的建筑空间（virtual architecture 与 virtual space），则建筑赖以维生的空间理论，将继埃及的量体空间、希腊罗马的几何空间、哥特式的神秘空间、文艺复兴后期到巴洛克的动态空间、现代与后现代空间之后，再出现建筑空间与都市空间以此的新空间理论。上述设计方法、设计思考、空间概念等三方面理论的共同发展，则有机会在建筑理论上建构一套全面的新的"主义"，继现代主义、后现代主义之后的所谓"数码主义"（digitalism），建筑的历史也将因而再向前推进。然而，如果建筑的设计方法论、思考理论、空间理论以及建筑历史都改变了的时候，数码建筑将不只是一个新的理论。

数码建筑因而可能是一个新时代。如果把所有的数码技术整合起来，视为对建筑设计具有全面性的影响的因素，并且接受现今人类历史发展的新时代便是数码电子时代这样的观念，则数码建筑是继史前时代、埃及、希腊与罗马（玛雅、东方、印度等）、早期基督教、拜占庭、哥特、文艺复兴、巴洛克与洛可可、新古典、现代与后现代等建筑历史上的一个新时代，或许称为数码建筑时代。值得注意的是，若数码建筑有机会形成一个"时代"，则它的影响力将不只是上述的大量的工具发展或各种理论的成熟而已。它将会全面改写建筑在产业、社会、文化方面的影响，以及引发在这些重大影响下所形成的另一种具有数码性的新价值体系（为什么好？）与新美学（什么是美的？）。

最后，数码建筑可以被定义为一场新革命。人类文明自混沌时期起，历经渔猎革命、农业革命、工业革命等重大变革，每次革命影响所及都改变了人类的思维模式与生活方式，建筑在人类文明的革命中显得极其渺小，只能随着革命所形成的新思维与新生活模式而做彻头彻尾的改变。人类文明是否会形成数码时代甚至引发一场数码革命，身在建筑领域的我们只能拭目以待。

上面谈到数码建筑的四种定位，我们除了从实质发展层面来衡量外，还可以用另一

The Inconsistency in Digital Architecture   It's not easy to predict the future of the digital architecture. Although, it's still progressing with the full speed. There are still inconsistency between the architectural education, society culture, and digital architecture.

The inconsistency between the professional and non-professional. As I mentioned previously, the revolution generally causes controversy. Even though many people only use computer to type and e-mail, they don't hesitate to criticize and question the digital architecture, In 1992, while I was attending a meeting in Boston, some people are still debating whether the computer can think independently, without knowing that the Deep Blue has already defeated the world chess champion.

The inconsistency between two exiting generation and digital generation. We define the generation every 10 years. Its not easy for the different generations to communicate with each other. The generation gap between the existing generation and the digital generation can be easily seen in the debate of architectural design process, presentation, space, volume, concept and aesthetics (a Taiwanese architectural student almost got flunk for using the computation presentation instead of

左下图:"数码自然" 电脑模拟全图
BELOW: "Digital Nature" computer simulated perspective

右下图:"数码自然" 2001 年 1 月 17 日
BELOW: "Digital Nature" Jan. 17th, 2001

个观察的指标——时间来判断。以一种新工具（数码工具）而言，它的发展持续期大约是数月，几个月后就会有另一种新数码工具产生。然而，一个新理论则可持续10年至30年（例如1970年～1980年的后现代理论及1990年的解构理论等），而新时代则有机会持续数十年至数百年，新革命则有数百年至数千年的时间（虽然时间延续的长度有越来越短的趋势）。因此，如果建筑在数码性的发展上仅有几年关键期，则它必然只是工具，若能持续发展10～30年，则它会是个新理论（这一点以哈佛、麻省理工、哥伦比亚等大学建筑系的眼光看来，它早就是新理论了），若能再继续发展并主导建筑领域数十年以上，就会有可能形成建筑的数码时代。当然，能主宰百年以上，就是新革命了。

数码建筑发展的断裂　由上面的分析看起来，虽然数码建筑到底可以发展到什么程度，我们无法预料，放眼过去10年以及未来，它发展的速度之快与影响之大，却是不争的事实。但是，就在电脑这样迅速且全面的影响着我们早已习以为常的"建筑"时，仔细观察，我们可以发现在可能的"世代交替"下，目前存在着下列几种建筑专业、教育、社会、文化上的断裂现象。

电脑常识与数码专业知识的断裂　如前面提到的，在任何一项新科技可能引发一场革命时，总有人会在对新科技缺乏全面性了解的时候便立刻反对并开始挑剔。在数码建筑的发展上，由于电脑实在太过普及，许多人会在没有吸收任何10年内有关数码建筑在设计上或理论上的"专业知识"的情况下，只是基于自己用电脑处理文书、电子邮件及上网的经验与"常识"便来评论数码建筑，很少有人乐意说他只熟悉文书与网络，而对数码建筑没有深入的了解。这个状况让我想起1992年在波士顿的一场会议中，在许多人辩论着电脑是否可能拥有思考能力甚至拥有创造力时，反对的人的电脑经验大都仅限于电脑打字的常识，而完全缺乏对人工智能的起码了解，甚至不知道电脑"深蓝"的思考能力已击败国际象棋冠军，更不知道电脑正在向创造性思维迈进。

tradition drawings and models). The gaps exist everywhere in the world. For instance, the digital generation uses computer to white, thinking the traditional writing time-consuming, while the other generation criticizes the writing in the e-mail as "the writing without quality". This kind of criticism sounds familiar, we can trace it back to seventy years ago, while the Chinese modern literature replaces the classical literature. The Renaissance people who writes in quill and brown ink may have the same criticism towards the stationery today.

The inconsistency between the architectural and non-architectural field. Focusing more on the arts, history and culture, architecture always reacts slowly for the technology revolution. For instance, the communication has been totally digitalized, the architects still merely use metal and glass to express the avant-guard architecture, without digitalizing the design process. Meanwhile, the architectural education hardly puts any effort to keep up with the rapid change of the digital technology. It's possible that the traditional architectural education is still based on the philosophical, social, historical and psychological theory. Some exceptions I know are MIT, Columbia, and ETH.

左下图:"数码自然" 2001 年 2 月 25 日
BELOW:"Digital Nature" Feb. 25th, 2001

右下图:"数码自然" 2001 年 3 月 26 日
BELOW:"Digital Nature" Mar. 26th, 2001

非电脑时代的人与数码时代的人的断裂　以往我们区分时代的方法是纪年，而且大约将10年分为一个年代，因而形成不同时代在想法与价值观上的"代沟"（generation gaps），相互不理解对方而且不易沟通。有了电脑之后，却形成了与年龄无必然关系的电脑时代与非电脑时代的人的代沟。以家电用品为例，好的电视与冰箱必须是坚固耐用，而好的电脑万万不能坚固耐用而且必须在3年内更新。电脑所带来的思维方式与价值观的断裂，经常出现在讨论数码建筑中对截然不同的设计方法、设计过程、设计表现手法、设计的空间与量体、设计美感与观念等等的辩论。（我曾听说有个大学的学生因用电脑表现设计图与模型，因此没有达到系里传统贴图和大模型的标准，而差一点未通过考试。）另外，除了在建筑上的时代落差之外，在一般生活的价值观上，也处处可发现非电脑时代与数码时代的断裂，例如，电脑时代抱怨传统书信的低效率与低功能，并认为数码是个必然的趋势。而非电脑时代的人除了无法想像竟有一群毛头小子利用网络就可赚数亿美金，也会将电子邮件视为"缺乏感情的书写"的元凶。但这样的批评，看起来是如此的似曾相识。现在我们所使用的"正式且优美"的白话文，在五四运动之前，难道不也被视为难登大雅之堂。而文艺复兴时代的文人们书写时，向来讲究水笔、咖啡色墨水，以及考究的纯棉纸张，以他们的习惯来看今日在笔、纸、墨水等"劣等质感"时，他们是否也会有同样不以为然的感叹。

建筑圈外与圈内的断裂　建筑是个注重艺术、历史、文化的学科，因此对于科技为主导的变革，经常反应较迟缓而且有时会有反变革的情况。建筑圈外，由于电脑科技的发展，许多学科已发生"本质"上的革命（例如通讯已完全数码化），而建筑圈内谈到科技，许多时候仍停在运用玻璃、金属材料来表达科技"意象"，而不作任何追赶数码技术发展的尝试。此外，建筑教育本该负责为职业界介绍新知识并发展新理论，但可能有些建筑教育界习惯于哲学性理论以百年至千年为变化单位、社会性理论以十年至百年为单位、历史与心理学理论至少也有十年以上的持续性，而忽略了电脑学科应在理论与技术发展，也和它的产品一样，有迅速更新的本质，所以必须2～3年就有理论或技术性的成长

The inconsistency between digital technology and human science. Emphasizing its philosophical, historical, social and cultural context, architecture is always the essential part of the human science. Many people still worry the overusing of computers. However, they forget that the new technology always causes new culture, like the printing and the Effel Tower. (When the Effel Tower was built, there was even the traditional artist who committed suicide to protest.) I sincerely believe, the computation will enhance the cultural development a lot, same as the architecture.

We Are Defining Architecture in the Digital Age   In the architectural and arts history, new theory is destined to be "criticized" by the contemporary society. Only few of these new theories can be confirmed, producing master pieces. The vision of digital architecture also needs both the personal creativity and the social appreciation.

左下图:"数码自然" 2001 年 3 月 28 日
BELOW:"Digital Nature" Mar. 28th, 2001

右下图:"数码自然" 2001 年 4 月 02 日
BELOW:"Digital Nature" Apr. 2nd, 2001

投入建筑设计，而跟着改变课程的内容与方向，如近年来麻省理工学院、哥伦比亚大学、瑞士联邦工业大学(ETH)等校建筑课程的新发展皆是如此。

数码科技与人文的断裂　建筑一直是人文学科中重要的一支，也十分强调建筑中所蕴含的人文精神（哲学、历史、社会、文化等）。然而，任何新科技发展的初期，因为其所可能造成的巨大变革，经常被卫道者以它是反文化、反艺术的东西等立场，大加批判，而且语重心长地劝人千万要小心它的大量发展所会引发的后果。然而他们却忘了，新科技的发展一直是人类文化成长的关键因素，就像印刷术是科技还是文化？艾菲尔铁塔的钢铁是科技同时是文化？（当艾菲尔先生看到工业革命的契机而坚持建造铁塔时，反对声浪大到有传统艺术家跳楼进行死谏。以这个角度看来，数码时代所遇到的反对声音比起工业革命时，真是小巫见大巫。这是否可算是人类在文明发展上的渐趋成熟？）我十分相信，电脑的发展就是人类文化发展的一大步，建筑也将在这个文化跳跃中大幅演进。

我们正在为数码建筑下定义　从艺术与建筑历史的发展上我们可以清楚看到，任何一项新学说、新风格、新创作以及跨时代的新思维，除了需要创作者的全心投入之外，这些新想法还必须受到当时的社会文化力量的批判与严格评鉴。严格筛选之后，大部分的新想法只被评价为新点子，而只有极少数的新想法会通过层层考验而得到最终的肯定，有机会在历史上留名成为新理论与新创作。所以，创作者个人的创作力与社会文化的鉴赏力，两种力量的相互配合才能完成艺术与建筑的跨时代发展。

目前，有许许多多的人正在世界各地日以继夜全力探究数码观念对建筑极限的冲击，换句话说，他们正想为建筑在数码时代中书写历史，为20～30年后的数码建筑下定义。2000年在中国台湾地区利用网络举行的远东数码建筑奖(Far East International Digital Architectural Design Award, FEIDAD Award)，就是尝试在全球数码建筑发展的关键性初期，为各国建筑师与年轻设计者在数码建筑的各种发展的可能性上，提供一个全面的交流、讨论、观摩、竞争的场所与机会，并通过颁奖予以肯定，与创作者在社会文化的层

Many people are making their most efforts to define the digital architecture. They are writing the history of digital architecture. The goal of this FEIDAD Award is to gather the wisdom of the young architects. In this competition, we have 204 works from 43 countries, more than half of the competitors are architects One of the winners is commissioned as the lecturer of Oregon University, while the other one is hired by the UN studio as the designer. Many works have become the teaching material as well. All these prove the importance of the digital architecture. In addition to managing the digital architecture award, the architecture program at National Chiao Tung University has recently explored a practical project "Digital Nature" in Taipei, in order to deepen the discussion of digitality of architecture. (Digital Nature, the reception hall of Bcom Electronic, INC. Designers: Yu-tung Aleppo Liu, Yuan-jung Lee, Sheng-cheng Shih, Shih-cheng Huang, Kuo-shang Huang, Chia-lun Chang, Te Lai, and Yang-chen Fan)

Before closing, I would like to express my gratitude to all the people whose efforts made this award possiblemy colleagues at Far Eastern Memorial Foundation and the review jury, my students at NCTU, press reporters and all the participants. [Translated by Moonlim Jau]

左下图:"数码自然" 2001 年 4 月 09 日
BELOW:"Digital Nature" Apr. 9th, 2001

右下图:完工后的全景
BELOW:The wide view when finished

参考书目

Liu, Y. T. (刘育东) 2000. The evolving concept of space: From Hsinchu Museum of Arts to Digital City Art Center. ACADIA Quarterly 19, no. 4.

Liu, Y. T. (刘育东) 1998. Restructuring Shapes: Design cognition and computation. Ann Arbor, Michigan: Proctor.

刘育东，1997.《建筑在电脑时代中何去何从》(Where should architecture go in the computer era)? 建筑杂志(Dialogue) no. 9: 31-33

The 2000 FEIDAD Award. http://vlab.iaa.nctu.edu.tw/Archi/FarEasternPrize/DigitalEnglish/

面上互动，共同为数码时代的建筑逐步下定义。在2000年的奖项中，来自43个国家和地区的204件作品有一半以上的参赛者是建筑师。作品的平均水准深受全球各国的肯定，得奖的前五名设计师均在事后接到邀请前往美国俄勒冈大学建筑系担任专任教师，也接到许多如UN Studio等知名事务所的邀请担任设计师。另外，前100名的作品已纷纷被各校建筑系选用为课程的讨论素材，这都说明了数码建筑在社会文化中的前景。中国台湾除了主办数码建筑奖以外，交通大学建筑研究所也为了能在建筑的社会文化发展上，有机会讨论数码建筑设计与施工实务的问题，自2000年11月至2001年4月推出数码空间设计—"数码自然"。（中国台北公信电子公司大厅设计，参与人员：刘育东、李元荣、施胜诚、黄士诚、黄国贤、张嘉伦、赖德、范扬铮。）

因此，本书将2000年远东数码建筑奖前100名作品集结出书，目的是希望记录2000年全球各地对"数码建筑"的不同角度的定义与看法，为数码建筑发展史作一次全面的整理。其中，我们看得出来"空间概念的演化"、"参数智能的设计"、"空间与形体的解放"、"数码类型建构"与"数码媒体的呈现"等方向，是目前数码建筑发展上的重点。本书能顺利出版，最重要的是中国远东集团总裁徐序东先生以及基金会执行长黄茂德先生对建筑的远见，乐意支持不会立刻看到成果但必然影响深远的远东数码建筑奖。另外，包含Peter Eisenman、Bernard Tschumi、William Mitchell、Reed Kroloff 等大师级国际专业人士在内的35位全球评审的热情支持，才能吸引204件好作品的参赛。另外，我要向提供作品的84位"数码建筑师们"（"digital architects"）致谢，以及特别感谢一年前与我们一起工作至今的朋友：基金会的胡湘君与陈明仪小姐、交大的黄英修、陈怡婷、施胜诚、李惠琳同学、《Dialogue》的金光裕总编辑、孙淑兰、洪士尧编辑等正是他们的长期投入，再加上丁荣生、张伯顺、赖素铃、谢慧青、潘彦蓉、郭孟君、张肇麟、林芳怡、林柏年、黄湘娟、吴光庭等朋友支持协助，才使得远东数码奖与本书能有今天的一些成果。

Juror List of 2000 Far Eastern International Digital Architectural Design Award
2000 年远东国际数码建筑设计奖评审名单

委员会主席 / Chairman of Committee：
城仲模先生 Chung-mo Cheng

总召集人 / General Convoker：
汉宝德先生 Pao-teh Han

召集人 / Convoker：
刘育东先生 Yu-tung Liu

The Preliminary and Second Jurors / 初选与复选评审

刘育东先生 Yu-tung Liu
中国台湾交通大学人文社会学院副院长 / National Chiao-Tung University

王泽先生 Joseph Wang （依姓氏笔划）
中国台湾铭传大学建筑系教授 / Ming Chuan University

吴光庭先生 Kwang-ting Wu
中国台湾淡江大学建筑系教授 / Tamkang University

季铁男先生 Ti-nan Chi
中国台湾乙建联合建筑师事务所主持人 / Chi's Workshop

陈珍诚先生 Chen-cheng Chen
中国台湾淡江大学建筑系教授 / Tamkang University

张基义先生 Chi-yi Chang
中国台湾交通大学建筑研究所教授 / Chiao-Tung University

黄声远先生 Sheng-yuan Hwang
黄声远建筑师事务所主持人 / Sheng-Yuan Hwang Architects & Planners

曾成德先生 C. David Tseng
中国台湾东海大学建筑系教授 / Tunghai University

John Gero
University of Sydney, Australia

Bob Martens
Technical University Vienna, Austria

Jerzy Wojtowicz
University of British Columbia, Canada

Tomas Kvan
University of Hong Kong, Hong Kong

Jin-Yen Tsou
Chinese University of Hong Kong, Hong Kong

Mitsuo Morozumi
Kumamoto University, Japan

Tsuyoshi Sasada
Osaka University, Japan

Shigeyuki Yamaguchi
Kyoto Insitute of technology, Japan

Alexander Koutamanis
Technical University Delft, Netherlands

Gerhard Schmitt
ETH Zurich, Switzerland

Tom Maver
University of Strathclyde, UK

Mark J. Clayton
Texas A&M University, USA

Nancy Cheng
University of Oregon, USA

Jeffrey Huang
Harvard University, USA

Brian Johnson
University of Washington, USA

Branko Kolarevic
University of Pennsylvania, USA

Malcolm McCullough
Carnegie-Mellon University, USA

William Porter
MIT, USA

Rick Rosson
Eisenman Architects, New York, USA

The Final Jurors / 决选评审

汉宝德先生 Pao-teh Han
中国台湾文化艺术基金会董事长、前中国台南艺术学院村 / Chairman, Arts & Cultural Foundation; Ex-principal, Tainan College of the Arts

刘育东先生 Yu-tung Liu
中国台湾交通大学人文社会学院副院长 / Vice Presider College of Humanities and Social Sciences, Chiao-Tung University

（ 以下依姓氏笔划 ）

姚仁喜先生 Kris Yao
大元建筑及设计事务所建筑师 / Principal, Artech Inc.

徐旭平先生 Peter Hsu
中国远东集团资讯长 / CIO, Far Eastern Group

Peter Eisenman
President, Eisenman Architects, P.C., New York

Reed Kroloff
Editor-in-Chief, Architecture magazine

William Mitchell
Dean, School of Architecture and Planning, MIT

Bernard Tschumi
Dean, Graduate School of Architecture Planning and Preservation, Columbia University

2000 FEIDAD AWARD Activities
2000 年远东国际数码建筑设计奖活动纪实

Launching —活动起跑

Election and Forum of FEIDAD AWARD —作品评选及数码建筑论坛

1.

2.

3.

1、2、3—The digital architecture forum which is discussing the situation of architecture in digital era, is well discussed by experts from different fields/
1、2、3—数码建筑论坛—对数码时代的建筑风貌的议题，邀集专家学者交流讨论

The Lectures of Architects from Different Countries / 国际建筑师讲座

ABOVE: The Architect: Peter Eisenman and Translator: Chung-sheng Tan /
上图:国际建筑大师 Peter Eisenman 以及翻译谭重生先生

ABOVE: The lecturing of Peter Eisenman/ 上图: Peter Eisenman 演讲盛况

ABOVE: The co-work of the academy, the politics and the corporations provides the opportunity to new development in architecture. Hsin-chu Mayor Jeng-jiang Tsai (L1)、 Architecture Peter Eisenman(L2)、 Headmast of NATU Jiun-yen Chang(M)、 Vice Persident of College of Humanities and Social Science, Yu-tung Liu(R2), and CEO of Far Eastern Memorial Foundation Morton Huang(R) / 上图: 学界、政界与商界的合作同盟，提供了中国台湾地区建筑发展新的契机。左起新竹市蔡仁坚市长、国际建筑大师 Peter Eisenman(左二)、中国台湾交大张俊彦校长（中）、中国台湾交大人文社会学院刘育东副院长（右二）及基金会黄茂德执行长（右）

The Final Election and Award Ceremony of FEIDAD AWARD / 决选阶段及颁奖典礼

ABOVE: The General Convoker of FEDAD review group, Mr. Po-teh Hang (L) and the Chairman of FEDAD review group, Mr. Chung-mo Chen (R).
上图：远东建筑奖评审团总召集人汉宝德先生(左)及评审团主席城仲模先生(右)。

ABOVE: Mr · Jiun-hsiung Chang (R) and Douglas Tong Hsu, the Chairman of Far Eastern Group (L)./ 上图:张俊雄先生((右)及远东企业集团徐旭东董事长(左)。

ABOVE: Jiun-hsiung Chang (R) right presents Outstanding Award to Danijela Pilic and Barbara Leyendecker, German Graduate School students.
上图：张俊雄先生((右)颁发首奖奖金及奖品予德国硕士班学生 Danijela Pilic 及 Barbara Leyendecker。

ABOVE: Hsin-chu Mayor Jeng-jiang Tsai (R) delivers Merit Award to Kuo-chien Shen, NATU Architecture Graduate School student (L)./ 上图：新竹市蔡仁坚市长(右)颁发佳作奖奖金与奖品予中国台湾交大建筑研究所学生沈国建先生(左)。

Recognition from Abroad — Teaching Invitation from the University of Oregon, U.S.A.
来自国际的肯定—美国俄勒冈大学任职邀请函

美国俄勒冈大学建筑系近年来积极发展数码建筑，该校肯定本数码建筑奖的品质，在征得评审召集人的同意下，邀请前五名作品的设计师前往担任建筑设计专任讲师，并极积鼓励入围前四十名设计师在该校任职教学。下列电子邮件即为俄勒冈大学邀请四位佳作奖得主之一，中国台湾交通大学建筑研究所应届硕士毕业生沈国健先生的邀请信函。

Dear Kuo-Chien Shen,

Congratulations on your success in the FEIDAD competition. As one of the preliminary jurors, I was impressed by the high level of submissions.

Your work impressed us with both architectural concepts and multimedia presentation. The high-energy presentation was very exciting to explore.

If you would consider a teaching position in the United States, we would be very happy to receive your application. In our school of Architecture and Allied Arts, your wide range of talents would be strongly appreciated, as the school as just initiated a new Multimedia Design program. As with all our faculty, you would teach both your specialty classes plus design studios. The quarter system allows flexibility in teaching.

For technical support, we have very service oriented people both at the University Computing Center level and for our own School of Architecture and Allied Arts (2 fulltime staff supervising a team of students). The Computing Center produces a newsletter and co-sponsors courses on basic skills and the Library has a Faculty Instructional Technology Center.

The campus network is on the fast Internet-protocol so that it has been called one of the most wired campuses. The university teaching labs are well-equipped and well managed. All our studios and most of our classrooms and pinup rooms are wired, with about half of our students bringing their computers to use in school. Some information is available below.

Sincerely,
Nancy Cheng
Computing Center / School of Architecture and Allied ARts / Department of Architecture / Multimedia Design Program

Top Forty Contestants' Works / 前四十名作品一览表

No.	Name	Nation	Occupation	Title	Theme and Description
1	Danijela Pilic Barbara Leyendecker	Germany	Student (Graduate)	i_map	i_map is a media based exhibition program that deals with the exhibition of information. The focus is on the symbiosis of material and immaterial space: the exhibition, the software, cannot be experienced without a built museum, the appropriate hardware. nevertheless i_map is not a classical architecture project but rather a concept that uses the interaction between reality and our real world of matter in order to create architecture.
2	Aaron Cohen	USA	Licenced Architect	Interface	Our society is breathing out into space, real physical architectural space. At the same time people use the web, electronics, artificial complex systems, and sensory spaces have started introducing new ways of seeing. We live the duality of digital vs. physical identity. We redefine the concepts of scale, of the original and the copy, of time and space, even their very own senses. The project interface is a research of ways for co-existence and experimentation on these new emerging issues, \ leading to a revisited materiality and a new architecture.
3	A. Scott Howe	USA	Licenced Architect	The Digital Condominium	Proposed is a Digital Condominium that is designed in cyberspace, uses online factories to manufacture its kit-of-parts, is constructed with robotic building systems, and is managed during occupancy by its own virtual self. The Digital Condominium was conceived to be a demonstration project which showcases a new construction concept that calls for the establishment of a site factory. In the site factory concept, traditional "final line" construction is replaced by many parallel manufacturing and assembly processes in the "assembly line" method. This includes the assembly of building components on the site using robots and automated construction equipment.
4	Birger Sevaldson Phu Duong	Norway USA	Professional Designer	Ambient Amplifiers	The goal is to apply time-based modeling techniques to generate architecture at an urban scale. These techniques are used to determine principles to inform the development of form and program. The islands are local amplifiers that help to interweave three different aspects of the Toyen Park (Science, Art, & Leisure) They share different modes of finishedness from footprints, foundations and space frames to enclosures.
5	Kuo-chien Shen	China	Student (Graduate)	Dynaform	The volume of this building, by defining the functions and the circulation relationships of environment as dynamic forces, new type of interaction of each space are produced.
6	Bernhard Franken	Germany	Licenced Architect	Parametric Design	It developed unique design approach-parametric design/digital continuity. Parametric design can create a complexity of form which is experienced as a spatial quality.
7	Michel Hsiung	USA	Professional Designer	Architectural Typology	The standard architectural systems can create new typologies when they meet and integrate. The interaction of the architectural system acts as the generator of the network system, a simple system that can evolve to a complex one. My investigation ranges from pipes, slab, truss and skins.

No.	Name	Nation	Occupation	Title	Theme and Description
8	Adrien Raoul Hyoungjin Cho Remi Feghali	France	Undergraduate Student	iNSTANT eGO	Plugged to our clothing, iNSTANT eGO is primarily a cluster of intelligent tissue folded over, waiting to be unfolded. When released, the interface unfolds, subjective time is trapped in our personal space, creating an ambiguous space-time relationship where other people's time is only subject to our personal intimate time.
9	Maarten Van Breman	Netherland	Licenced Architect	Living in the City	The aim of this competition is to propose ways of reducing economic and social isolation and creating a better quality of life for local people.
10	Agnes Zwara	Poland	Licenced Architect	Beyond the year 2050 - Definitions of Micro- and Macro-space Based on The Project of Manned Research Space Craft	This project is to look for the most consistently and constantly functioning systems of formations. This attitude leads to the opportunity for developing a qualitatively new form unassailable- of the architecture, which in itself is a complete and self-sufficient item. From the other hand, being the component in the aspect of Universal macro-space. It shows an activity system based on the common technological progress aspect, and the ship craft being the individual.
11	Philippe Luc Barman	USA	Licenced Architect	V-Mall	The purpose of the v-mall project was to design a center for retail and entertainment in the middle of downtown Los Angeles. The v-mall had to deal with its multilayered context as well as to create its own urban identity. The design process started with the research of the radical movement of a snowboder during an airborne performance, thus, the methodology called "radical-o-meter" was developed. It was applied to test the v-mall in contextual setting of several potential building sites.
12	John Barnett	USA	Student (Undergraduate)	music in architecture	The product presented the idea of representing sound visually and graphically. Some of these ideas are presented in the form of "building and space "while others represent music and sound through the use of solid geometry. And the second crucial item was the definition of architecture, rather than investigating architecture directly. It tries to explore the components such as geometry, lights, scale, movement, etc......
13	Laura Cantarella	Italy	Licenced Architect	Borderline	This is a suggestion about the city at this borderline time, the polycentric city, the multiethnic city, the dance city, the communication city, the hypertext city, the democratic city, the ant monumental city, the virtual city, the information city, the surface city, the media city, the interactive city, the nomadic city, the transition city, the shelter city, the political city, the people city, the borderline city.
14	Didier Castelli	Switzerland	Licenced Architect	Stratapolis — third millenium cities, 4 hypothesis	With the beginning of the third millennium, the City which didn't completely sustain the mutations of the society evolutions, can't develop based upon the models of the second millennium. The City has to morph into another entity. 4 hypotheses about 4 themes of development are suggested/Urban : expansion, structure, communications, spaces.
15	Eugene T. H. Cheah	Australia	Licenced Architect	Media Thek Melbourne	The project explores such ideas as hypertext to organise pedestrian circulation, spatial conditions of digital environments, and architecture as an interface for engaging with digital processes.

No.	Name	Nation	Occupation	Title	Theme and Description
16	Huang-ming Chen	USA	Professional Designer	A Building Which Exhibits Itself	Begin with a sequence of incremental and additive collage/physical modeling exercises. An interpretive analysis of honorific architectural exemplars reveals underlying spatial strategies. A unique formal vocabulary is derived from the analysis. The analytic methods developed condition the synthetic methods of the project. THE BUILDING, WHICH EXHIBITS ITSELF, is a prototype demonstrating variations of the formal system; the transformations and generations of each variation; and the presentation of various transformational processes.
17	Land Design Studio	UK	Professional Designer	Play Zone, Millennium Dome	The Play Zone is one of 14 themed zones set within the Millennium Dome at Greenwich. The content of this pavilion was internationally curated and presented as a showcase of new and future ways to play and interact, presented as 18 individual digital 'games' from digital media artist. Employed 3D-computer modeling as an integral design tool at every stage of the project.
18	Santosh Dhamat	USA	student (Graduate)	Livable wall	The belief in a 'livable wall' emerged from exploring the possibilities that a wall can transform itself into various spaces. This project presents a model of an idea that transforms flexibility of digital culture in the physical space. The global mesh of connectivity plays with time and space to create an asynchronous media of interchange. Here a wall is a representative of built environment, which is in fact just a medium to render an idea. This wall can be a floor, ceiling, a piece of furniture or an object in space. This object is any instance in constantly transforming 'form'.
19	Stephen Donald	UK	Licenced architect	Future City	We make our networks and our networks make us.
20	Shawn Douphner Fred Holt Paxton Sheldahl	USA	Student (Undergraduate)	Analog Digital Process [adp]	Analog digital process: The physical product of the analog become digitally compressed; a flattening of spatiality; analysis, selection and interpretation follow. The digital becomes the physical once more, a perpetual process. Six phases: Analog space capture; Digital space capture; Analog-space schemas; Digital templates; Digital beliefs; Digital spatial manipulation device; Digital spatial insertion.
21	Kenneth Ho	USA	Professional Designer	Hyper-sequenced Territories Cinemap	The cine-mappings outlines a methodology for describing event territories through hyper-sequences. Cine-This case refers to the cinematic or the mode of representation of events in a moving sequence of images , and establishes the thesis interest in sequencing. Mapping-describes the act of marking or identifying territories, staking out the range of an event.
22	William Hsien Koon-Wwai Wong	USA	Student (Undergraduate)	Martial- Art Headquarter	The concept of this project is to achieve the balance between the body, mind and spirit, which are the three essences in takewondo. As the concept evolves, the digital tools allow the image of kicking-motion to transform into a rough shape of the building envelope.
23	Shi-cheng Huang	China	Student (Graduate)	Parasite Theather	"Parasite" used as--parasitic space lives in an existent architecture, takes it as its base, and gets tentative or permanent co-exist with the original architecture, the host, by expanding, invading and conjunction.

续表

No.	Name	Nation	Occupation	Title	Theme and Description
24	Lgor Kebel	Slovenia	Professional Designer	Just-in-time Infrastructure Prototype	Just-in-Time prototype development tries to answer to four questions throughout the various degrees: Domestic, working, public and mobile institutions are merged into one (4 into 1).The evolution of the conceptual design and programming engineering (4 into 1) of such a prototype is Just-in-Time presented in these 3 categories: global analysis of infrastructural networks, from time into space, infrastructural bodywork.
25	Andreas Karaiskos	UK	Licenced Architect	Body is Environment	The visual penetration and authority of a surgeon/doctor gynecologist, all involve a resignation to authority and power on the part of the individual who has little or no choice-it is not always possible to operate on oneself. This is the social convention that will be opposed through the enhancement of the body and its renovation through this design process. The nature of boundaries will be explored and through digital penetration of the body and space, architecture form will evolve.
26	Ning Gu Mary Lou Maher	Slovenia	Professional Designer	Architectural Design of a Virtual Campus	ETH World as the third campus in the virtual domain for ETH Zurich. ETH World will become a link between virtual domain and physical domain The design concept includes a 3D Virtual World as a place for building the community through meeting people and as a 3D space for accessing the variety of information repositories.
27	Fumio Matsumoto Shohei Matsukawa Akira Wakita	Japan	Licenced Architect Student	Infotube	This Tube is covered with lots of rectangular "cells," a minimum unit of consumer information, on which various images and texts are displayed. Cells can be freely linked to each other. This project proposes a new type of cyberspace on the Internet which has a close relation to the shopping street in the real world.
28	Marc Maurer	Netherland	Licenced Architect	P2001 - Graffiti Related Architecture	Our main interest lays in the exploration of the borderlines and areas between the design fields. The theme of the border and the potential tension of it within the architectural composition are other aspects of our interest. We hooked up with two of the most important Dutch graffiti-artists in spatial typographic design and started to design graffiti-related architecture together in a design team. The result is a new kind of media architecture that has a deconstructive aspect to it.
29	Boris Brorman Claus Peder Pedersen Claudia Antonia Morales Morten Daugaard Peter Hemmersam Tom Nielsen Lars F.G. Bendrup Thomas Larsen	Denmark	Licenced Architect	H-city	It tries to take advantage of the existing cultural and architectural values in relation to that cultural landscape we have chosen to label nature.
30	Dimitris Rotsios George Bakoulis	Greece	Licenced Architect	Future Vision Housing	The boundary can be grasped as the point of contact between the individual and collective intelligence, the active and passive parts of an interaction, of which the habitation is the epicenter. The point that justly can downgrade "fluid object" to a "household". The point where all natural needs coexist with all the perceived experiences.
31	Ori Scialom	Israel	Licenced Architect	A Terminal at the Haifa Port	Architecture today is facing new challenges. This project is trying to deal with the relations between Form and Information. Data regarding the quantities in relation to time was collected. This data was put into diagrams showing its changes during day and week cycles.

No.	Name	Nation	Occupation	Title	Theme and Description
32	Silvia Simoni Lorenzo Noe Luigi Fregoni Marco Valentino	Italy	Licenced Architect	Slotmachine	"Slotmachine" is an evolution of "Layers", a project presented to the International Venice Architecture Exhibition for the 7th Biennale.
33	Peter Stec	Slovakia	Student (Graduate)	Memory Constellations	Memory Constellations emerge from personal behavior on the web Data entities are connected into spatial structures of information. This proposal It develops methods of spatial self-organization of information combined with automatic creation of a communication interface. In development, it is now only a mockup of the real engine and may look hierarchical.
34	Tristan d'Estree Sterk	Australia Canada Holland	Licenced Architect	Idea Cloud	The metaphor of a cloud has been chosen to propel this project. A cloud is a field. A field is a complicated, whimsical, ephemeral, system of balances. It is an envelope or zone in which the possibility of "cloud" emerges. They come to being out of environmental conditions at any time, in any place wherever the circumstance provides. They are whimsical, pushing at the limits of their own existence, spinning and weaving and soaring and storming through space.
35	JMK Architects	USA	Licenced Architect	Library of The Third Millennium	Antonello de Messina's painting, "St. Jerome in his Study" served as a touchstone for the project, particularly for its embodiment of qualities in a library that should never change, regardless of how its contents do.
36	Carl R. Tully Dace A. Campbell Mark Farrelly Susan Campbell Bruce Campbell Taylor Simpson	USA	Licenced Architect	NPL	The National Public Library (NPL) will provide free remote and local access to digital media. This NPL will be a "cybrid" institution, with virtual and physical components: VIRTUAL LIBRARY, INFORMATION KIOSKS, PHYSICAL BRANCH LIBRARIES.
37	Shih Weng	China	Student (Graduate)	Dynamic Linear Membrane Infrastructure for Festival	The linear membrane as the medium of time.The linear membrane is a performing stage of image, shadow and light.The linear membrane is not only a container of information, it becomes information itself. The information becomes an architectural material.
38	Yi-yen Wu	China	Student (Graduate)	Liberal State	"Liberal State" is a kind of author's spiritual reaction to the distrust and unstability of the virtual world when he lives in the real world.
39	Chiafang Wu Stephen Roe	China Ireland	Professional Designer	Animate Building	As a starting point for realizing the digital architecture we envision an "integrated surface," incorporating structural, environmental and information technologies in a smart construction which responds intelligently to influences both inside and outside the building acting as an interface between the two,in much the same way that our own skin mediates between our bodies and the environment.
40	David Wylie	UK	Licenced Architect	Wylie Associates Web Site	Our web site is our means of demonstrating our approach to a world audience. Conceived as a virtual office, it offers an interface our visitors can identify with, allowing a set of complex ideas to be laid out clearly and imaginatively.

Juror's Comments of the Top 8 Works / 前八名作品评审结语

i_map

Outstanding Award / 杰出奖
Barbara Leyendecker + Danijela Pilic
Germany / 德国

This is a well-executed project at every level. In this project, virtual space matches with real space; this exploration is the most important form of research today. The idea of integrating hardware, software, and roomware to create a museum experience is also a good one. In addition, the presentation is clear, with elegant graphics and a well-designed interface.

就每个层面而言这个方案完成得非常好。在本方案中虚拟空间与现实空间相配；这样的探索是当今最重要的研究形式之一。结合硬件、软件与空间设施创造出来的博物馆体验的想法也是很好的概念。再者，优雅的图表与设计良好的界面也呈现出很清楚的结

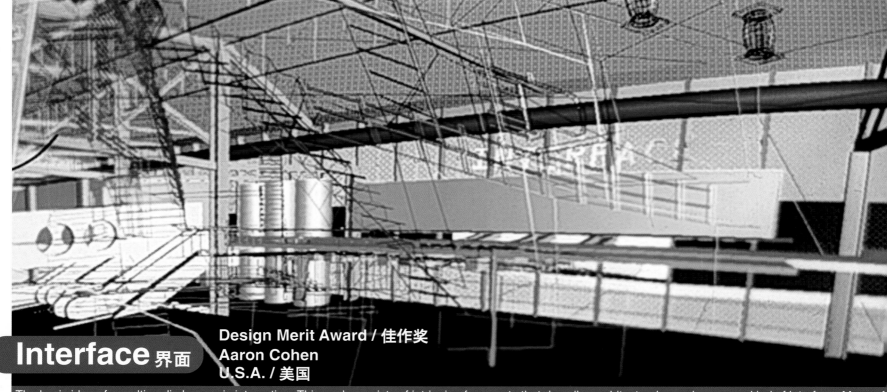

Interface 界面

Design Merit Award / 佳作奖
Aaron Cohen
U.S.A. / 美国

The basic idea of a multimedia hangar is interesting. This work consists of intriguing fragments that describe architectural experience as a kind of interface. Many of these fragments or presentations constitute essential aspects of what architecture is.

多媒体机棚的基本概念野蛮而有趣。本作品里含有将建筑经验描述为一种界面的复杂成分。这些成份或展现构成了建筑的必要元素。

Digital Condominium

数码公寓

Design Merit Award / 佳作奖
A. Scott Howe
U.S.A / 美国

An unsympathetic critic might dismiss this as an Archigram knockoff. Still, it is an ingenious mechanic-set which has been carefully developed and very clearly and completely presented. The idea of digital architecture is incorporated within this framework. This project also shows significant investments of time in the refinement and exploration of an idea as well as the associated web site.

一个没有同情心的评论者可能将本作品驳斥为"只廉价地卖弄建筑图像"。不过它仍是谨慎发展并清楚完整呈现出来的巧妙机械组合。数码建筑的概念包含在其框架之中。此作品同时表现出对于发展概念与相关网站本身独特的研究与精练。

Ambient Amplifiers 周边扩大器

Design Merit Award / 佳作奖
Birger Sevaldson + Phu Duong
Norway + U.S.A. / 挪威 ＋ 美国

A clear, intuitive interface presents this project to excellent advantage. And the idea of the "islands" is a provocative one illustrating a step-by-step process from analysis of issues to proposal of landscape configuration. The most convincing part in this project is the activator.

本作品因为清晰、直觉的界面呈现而占优势。而"岛"的大胆概念，则呈现出对景观表面配置提案分析的渐进的过程。此方案最令人信服的是"催化剂"的部分。

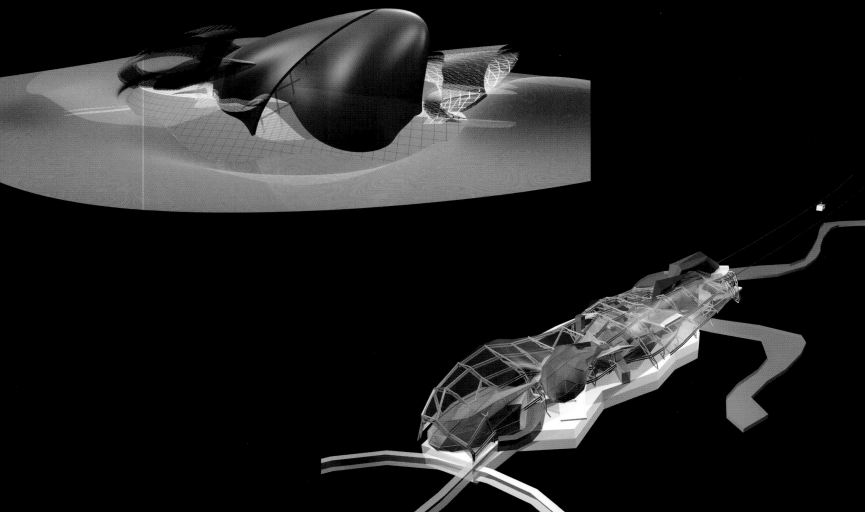

DynaForm 动态形体

Design Merit Award / 佳作奖
Kuo-chien Shen / 沈国健
Taiwan,China / 中国台湾地区

This project is full of energy and enthusiasm; it's an imaginative response to the problem of a cable car station. It is also presented with great verve. In addition, the attention to many architectural aspects of the project (skin, structure, section, program, etc.) is very appealing. There is still much to commend.

本方案充满活力与热忱，它用想像力解决了缆车车站的问题，并有着相当鲜活的呈现。其次，此方案的许多建筑观点(表层、结构、剖面和设计方案等)是动人的。本作品留下了许多讨论空间。

Parametic Design 设计典范

Special Prize for Digital Process
数码过程特别奖
Bernhard Franken
Germany / 德国

This project carried an idea all the way from conceptualization to CAD/CAM fabrication. It is a strong entry due to its realistic construction premises, also a convincing project, clearly and effectively presented: strong juxtaposition between the existing building and the wave-like interior. The tactile reality of the forms gives credence to the theory.

此方案从构想之初到利用电脑辅助设计建构都坚持着一个想法。就其建造条件的真实性而言它是个强有力的提案，也因为清楚有效的呈现手法使得它非常具有说服力：现有建筑与如波浪般的室内结构强烈的并置。形体触觉的真实性为其论点带来可信度。

iNSTANT eGO

瞬间的自我

**Special Prize for Digital Creativity /
数码创意特别奖
Adrien Raoul + Hyoungjin Chob + Remi Feghali
France / 法国**

This is a project that talks about changeable architecture and actually shows how it might happen. The "instant ego" bubble is a clever, poetic idea, and it is developed with verve and skill. Some of this project is a direct revival of the 1960s with the dream of an instant technology to build it: Archigram for the new millennium.

本方案探讨可变化的建筑，并且真实的呈现出变化发生的过程。"瞬间的自我"是个聪明、富有诗意的构想，并用活力与技巧来展开设计。它的某些部分，直接唤起了20世纪60年代以一种瞬间科技去构筑新千禧建筑图像的梦想。

Architecture Typology

建筑类型学

**Special Prize for Digital Analysis /
数码分析特别奖
Michel Hsiung
U.S.A. / 美国**

An interesting concept piece on building systems and typology. This project seems to be a good exploration of the contamination between structure and envelope. The entire process extends the classic understanding of architectural typology.

这是一个关于建筑系统与类型的有趣概念。此案似乎是对于结构与外壳之间一个不错的探索。整个过程扩展了我们对于建筑类型学的传统理解。

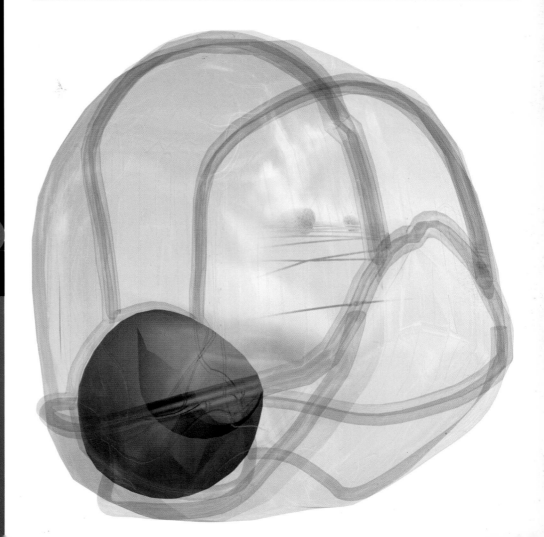

Liberation of Form

Liberation of Form and Space

Architects are like artists in that they possess limitless imagination during the design process. Architects, however, are constrained by their design media, and are unable to design spatial layouts that cannot be realized in the real world. They are also constrained by the materials used and construction methods. But with advances in materials and construction techniques, architects are able to make design breakthroughs. Architectural methods and construction materials and methods have developed from Egypt, to Greece, to the Middle Ages, to the Renaissance, to the Industrial Revolution, resulting in wider, taller, and lighter structures. Another great development was the use of models by Brunelleschi in the early Renaissance period, improving on the use of drawings that began in Egypt. This allowed Gaudi, Steiner and Uzon to design free-flowing structures. In 1990, Frank Gehry used computers to assist him to design free-forms, and then utilized CAD/CAM to quickly draw plans for the structures and actually build them. The digital design media and digital construction methods have liberated form and space in architecture, and have turned architecture into more of a pure art form than previously.

形体与空间的解放　　建筑师就像艺术家一样，在设计创作时经常会有无限的想像，但建筑师一方面受到设计媒体的限制，

画不出想像中的形体与空间，另一方面受到材料与建造技术的限制，即使画得出来也不见得建造得起来。建筑师一直在设计技

术与建筑材料的限制下从事设计，建筑也就随着设计技术与建筑材料的突破而产生时代性的变化。由古埃及、古希腊、中世

纪、文艺复兴时期、直到工业时代,建筑设计技术与建筑材料在历史上的发展，导致了建筑形体与空间的一次又一次解放

（跨距更宽、高度更高、量体更轻、曲度更大等等）。另一方面，文艺复兴早期意大利建筑师布鲁内莱斯基首度运用模型来思

考,弥补了自埃及以来画图的不足。这一创新,使得Gaudi、Steiner、Uzon设计出造型自由的建筑物。1990年Frank Gehry利用电脑

来设计他头脑中的自由造型,利用数码技术画图并最终将其建造起来。从此,数码设计媒体和数码建筑技术的革命,使得建筑

的形式与空间又产生了一次划时代的解放。建筑本身也因数码科技而变得更像是一种纯粹的艺术。

Ambient Amplifiers — The Logics of Uncertainty / 不确定性逻辑：周边扩大器

DESIGNER: Birger Sevaldson + Phu Duong

The relation between form and function, or an extended reading of function as program, has often been understood as either a one-way relation (Form Follows Function) or it has been challenged or denied. There is an undisputable casual relation between the need for appropriate tools to support human activity and the forms these tools develop over time. A deeper and inclusive reading of these processes includes cultural, political and psychological "functions" and the negotiation between these intentional or system-driven forces over time. The problem with these models for the generation of artefacts is not that they are wrong, but that they are incomplete. The production of artefacts needs to be seen as a process where the generation of form is influenced by action (function & program) as well as form influences action through hosting, triggering, modification and amplification of program. (Function Follows Form) These mutual form-generation processes take place at various (time-) scales. In everyday life we constantly adapt to existing conditions and in periods operate in an opportunistic mode, where our actions are shaped by the situation we face. On the other hand we constantly rearrange the form of our surroundings to fit our intentions and needs, from moving around chairs to redecorating, from product design to town planning. The role of the designer could be seen as a mediator and negotiator between those generative forces.

Ambient Amplifiers is a rehearsal on a real-life case, the Toyen Park in Oslo and its surroundings. The intention is to investigate the use of computer technology as engine for mediation of processes where form and program reconfigure and redefine each other in a mutual time based process. The design for such processes implies to prepare for a future myriad of potential events (actualities) that by far extend the imagination of the individual designers. Designing for certain programs is of limited value, because continuous reprogramming by the inhibitors will take place. In fact the intention is to reinforce such reprogramming and reforming. Therefore, the designer's anticipation and declared intentions can become an obstacle, since it easily produces solutions not only based on a reductive conception of what the program will be, but also informed by partly fixed design schemata. Instead the design needs to be open-ended towards the events that could take place. This open-ended state implies on one hand the resistance against typology, on the other an awareness of context.

The computer is used as an engine to produce dynamic generative diagrams, which serve to address the mentioned aspects. The dynamic generative diagram always feeds on contextual matter, but it deforms the fields of forms and forces in a topological animation, which produces series of intermediate steps, articulating and forming both spatial and temporal thresholds. The use of such techniques implies the introduction of an "untamed" set of information. The reason to do this is to resist the obvious "designed" solutions, which would only engage in a fixed scenario. The dynamic generative diagram is used as negotiation engines towards context as well as resistance buffers against fixed design schemata. The evaluation of the solutions is done through series of scenarios. Since the aim is to test the systems generic ability to adapt to unforeseen future states, extreme scenarios are most useful.

Location Toyen Park is a major urban park in the east of Oslo, Norway. It is situated between three groups of institutions: Science, Art and Leisure: [1]Science: Botanical garden, Paleontological Museum, Geological Museum and Zoological Museum; [2]Art: Munch Museum, Toyen Cultural Centre; [3]Leisure: The Toyen Bath, various public sport grounds.

The Park area forms a complex edge condition, which is formed by the following elements:
[1]Geological ridge running from north to south along the eastern edge of the park; [2]Urban fabric dissolving into suburban building structure; [3]Ethnic and social diversity between east and west side of site; [4]Transportation systems dividing the park into zones; [5] Fenced public institutions providing limited access to major parts of the site. The interventions reinforce the existing institutions, streets and their activities, rather than introducing new programs. Low activity level is not caused by a lack of richness in the park program but in a separation and lack of synergy between the programs.

Ambient Amplifiers invents and implements two slightly different types of diagrammatic machines: [1]Diagrammatic Machine 1 applies to the central field in the site. DM 1 injects a generic information set which negotiates a resultant with contextual information and design

形式与功能的关系经常被单方面定义为形式随功能，此过程可以说产生于包括政治、文化、心理等种种因素，以及这些制约力间的彼此妥协。这些模式本身并无对错，问题在于它们有所不足。人造物的产生同样牵涉到形式的种种寄托、击发、修正与扩大，加上动作的的繁复影响（机能随型）。每一天，我们都在随遇而安与因地制宜中挣扎；另一方面，我们又一再随心所欲改变着环境的形式，从搬搬椅子到重新装潢，从产品设计到城镇规划。

"周边扩大器"是奥斯陆 Toyen 公园与周边的一次实例预演。目的在于探究以计算机科技为动力，形式与程式二者同时交互定义、预知、作用的过程。这样一个为无数潜在因素做准备的设计过程，颇能启发设计师的想像，因为建筑计划有其局限，居民的新计划不断地发生。事实上本方案的目的就是强调这种变化。

电脑是上述动态式设计流程的动力工具。通过拓扑模拟原则，脉络式事件因此转化到力场上，产生了一连串接合时空的过渡性步骤。

"周边扩大器"包含两种不同的机制图，机制图一（DM1）应用于公园的中心地带，机制图二（DM2）适用于通往 Toyen 公园的走廊，[1]DM1 中和了原始脉络与设计企划 [2]DM2 将人潮加入 DM1 以维持活动的活力。这两种机制为以下元素提供了若干设计方向。

"公园岛群"如同本地扩大器，交织着公园里科学、艺术与休闲三种功能，借足迹、基础建设和空间骨架的围绕，共享着不同模式的完

AFFILIATION: OCEAN north
COUNTRY: Norway + U.S.A. / 挪威 + 美国

intentions. [2]Diagrammatic Machine 2 refers to the flow-feeding corridor of Toyengate. DM 2 is a derivative of contextual based information and design motives. DM2 feeds flow into DM1 to sustain vital activity.

The Diagrammatic machines produce the design underlay for the following elements: [1]Park Islands are introduced as point attractors to overcome the separation between the institutions and to create local attractions. [2]Street Scaffold is introduced to maintain a flow of people into the park. It accommodates new ways to inhabit the street by reconfiguring threshold relationships. [3]The Field creates opportunities to relate existing circulation systems for programming of events and activities. The organization of infrastructure over time results in increased flexibility.

The Park Islands are local amplifiers that help to interweave three different aspects of the Toyen Park: Science, Art, & Leisure. This is achieved through three sets of islands, each administrated by one of the three institutions. The sets are distributed in a way which brings the three institutions in close contact. Also on a administrative level the islands are meant to create synergies through a common comity.

The islands share different modes of finishedness from footprints, foundations and space frames to enclosures.

The Islands are not designed according to a specific program. Instead, the intention is to design programmability. This implies that there is a long range of possible uses for the different island types. Typical uses would be: [1]Footprints: Sports activities (vary according to surface), sun bathing, picknick. [2]Foundations: Art installations and sculptures, grill parties, sports activities. [3]Space Frame: Art

成。"岛"不是根据特定建筑计划而设计，而是设计空间计划弹性；这意味着不同的岛型可有广泛的用途，随着时间的推移，作用也可发生改变。

"街头鹰架"扮演着出入公园人潮的角色。它设下新边界条件，让现有建筑物共同参与；一系列置入、合并与扩张计划，产权、气候控制、公私运作等从此将有新的可能。

"场域"可作为不同动线间的连接，也连结不同的活动，这样的连结也增加了活动的弹性。［翻译 赵梦琳］

↓ Side view of island. ／ 岛的侧视

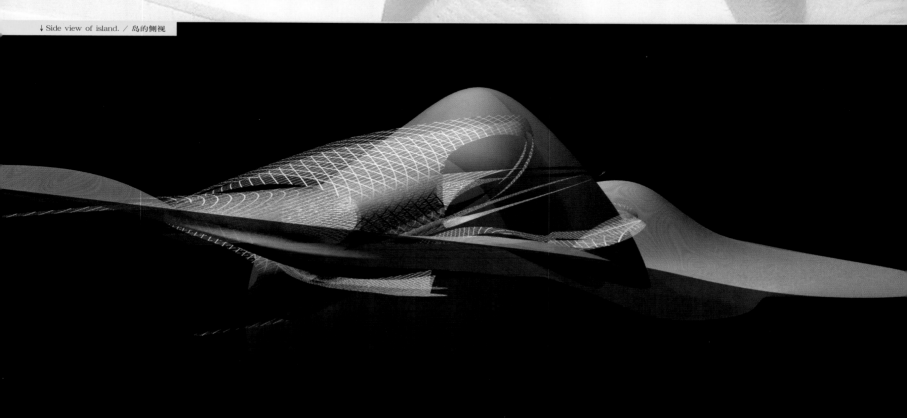

perfomances, temporal stands (tents) for parties or commercial events. [4]Enclosures: Exhibitions, gatherings, lectures, parties. New uses can be invented and changed over time.

Some islands operate as infrastructure by supplying water and electricity to providing communication access via broadband VAN. Islands are mapped with site relevant information including GPS modified mobile information.

The role of the Toyengate Street Scaffold is to negotiate activity flows to and from Toyen Park. The strategy is to reinforce and redistribute activity in the street to the Park Islands and Field. This intervention creates possibilities for rethinking the space of the street. The armature behaves as a local way-finder to signify Toyenpark. At the metropolitan scale it performs as an entry corridor visible from the adjacent street, Gronlandsleiret, which leads to the city centre.

The proposed structure transforms the urban corridor by interweaving additional programming that overlaps with local inhabitation patterns. The Toyengate Street Scaffold engages at the global scale by linking tourist flows from the city centre to the Munch Museum.

The street scaffold enables participation with existing buildings by setting up new threshold conditions. Through insertion, affiliation, and expansion relationships for program, property ownership, climate control, and public and private operation are given new potentials.

The Field interweaves three different transportation systems: [1]The street systems, mainly the Finmarksgate, which in its current state divides the park into an upper, and a lover area, which today is underdeveloped. The street system is redesigned in the proposal. [2]The network of existing pedestrian and bicycles paths. This system is largely left in its current state. [3]The system of hard surface. This is not a pure transportation system, but a crossover between pedestrian paths and surfaces for play and leisure, which blurs the borders between transportation systems and leisure surface. This system is introduced in the proposal.

The fence enclosing the Botanical Garden is another obstacle, reducing accessibility and dividing the site into hard-edged zones. Since the fence cannot be removed it is converted into a programmable device similar to the street. It can achieve a long range of diverse states of openness according to the time of day or if special events take place in the area.

The central area hosts a wide variation of activities that vary overtime. These variations take place at the scale of hours, days and seasons. The programmable street seeks to meet these changes in a more adaptable manner than the current situation. The proposal creates more fluid access to the central area. By dissolving the street and the fence into programmable devices, the demands for traffic flow, park life and time regulated needs are addressed.

↑ Sitemap with the three institutions, Art, Science and Leisure.
基地平面布置图

Art

Science

Science-park

Leisure

Leisure-park

Tøyengate

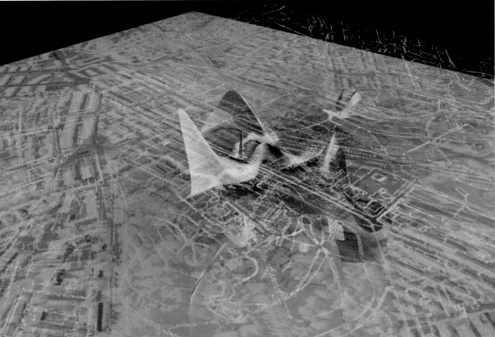

↑ Process study showing the interweaving of different modes of representation and the derivation of a generative diagrammatic shape. 不同模式的融合与动态图形发展过程

↓ Perspective view of island ／ 岛的透视

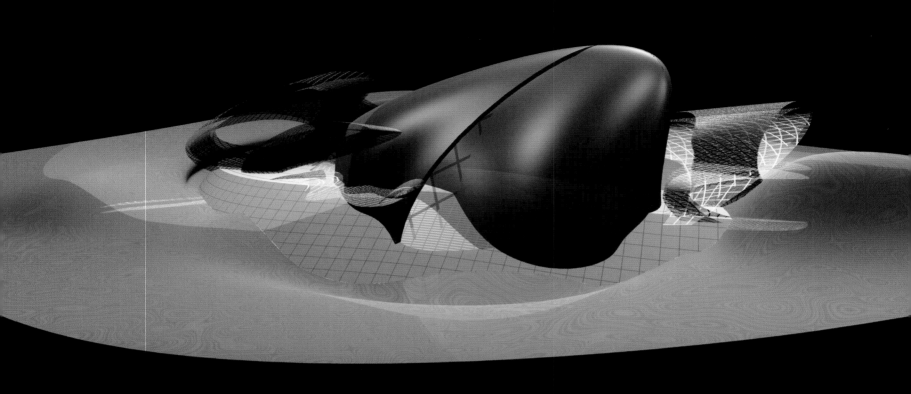

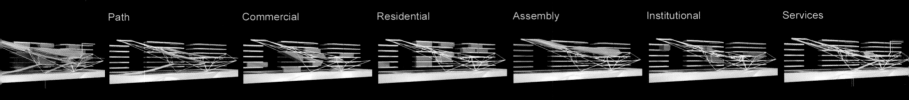

Path　　Commercial　　Residential　　Assembly　　Institutional　　Services

↑ Image series showing diagrams of how the street scaffold interrelates and amplifies diverse existing programs of the street. ／ 街头鹰架穿插联系及扩大街头各种活动的方式
↓ Polygonal set derived from the generative surfaces used as underlay for the Street Scaffold. ／ 多边性场景，源于街头鹰架以大量的表面为底
→Scenario of street scaffold intensity node. ／ 街头鹰架强度节点情境

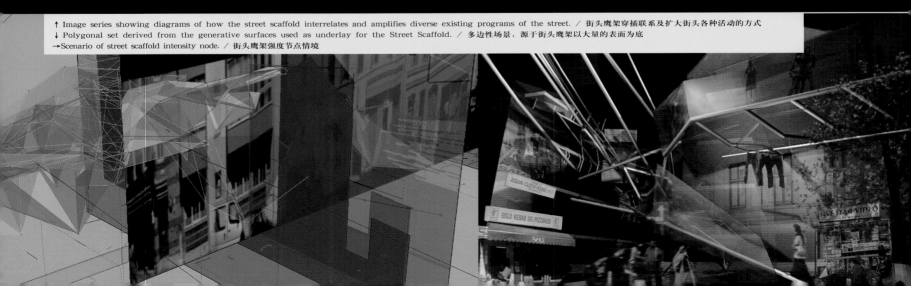

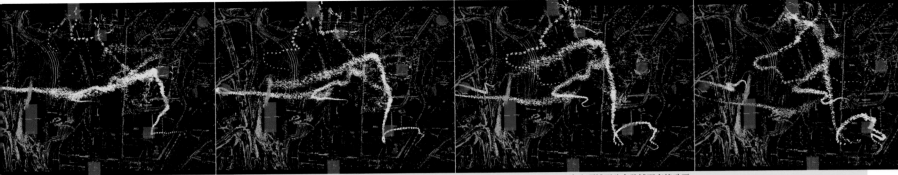

↑ Particle animation used to inform the distribution of the Islands, the "Hard Surface System" and several border conditions on the site. ／ 岛上硬铺面分布及铺面交接动画

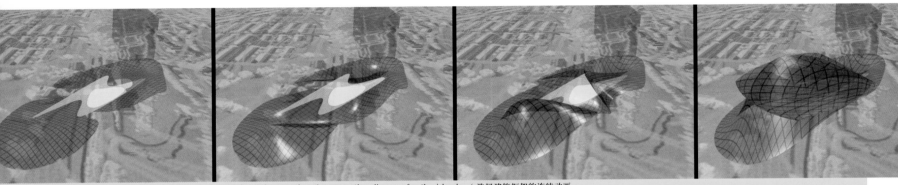

↑ Animation series showing selected frames of the deformation creating the generative diagram for the islands. ／ 选择建筑框架的连续动画

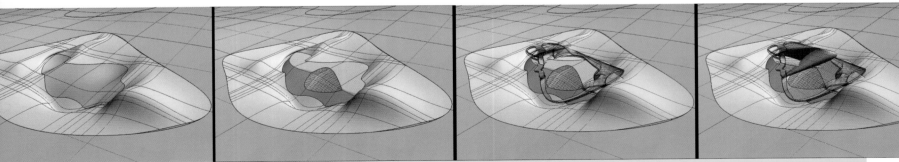

↑ Series showing the four states of unfinishedness:Footprint, Foundation, Frame, Core ／ 去掉表面覆盖的建筑：平面、基础、框架、核心

Parametric Design / 设计典范
DESIGNER: Bernhard Franken

Clean Energy Designing an exhibition is more a question of communication then architecture. Our design is not a container for information but communication itself. Starting point of the EXPO2000 Project for the BMW Group was consequently not a program or function but a briefing of a communication concept called Clean Energy. Clean Energy is the initiative of the BMW Group to use hydrogen as the fuel for the future of driving. The hydrogen technology is completely developed by BMW, the first cars are produced, but a network of gas stations and hydrogen produced by sustainable energy is not available. So it is rather a political then a technical question BMW wants to promote as part of the EXPO2000. Form, content and presentation was completely left to us to develop.

Site In a collaboration with the Deutsche Museum Verkehrszentrum the Clean Energy Exhibition is an official international Project of the EXPO2000. Since most of the former Messe is going to be replaced, so far the area is more resembling a wasteland. All around 3 protected old halls heavy construction work is going on. The halls themselves, build in 1908, are in a terrible condition. Since our Clean Energy exhibition is lasting for 5 month we decided to emphasise the temporary status of the project

The Wave The event theme "Clean Energy- driving with sun and water" itself is creating its appropriate form. Energy exists in different states. The sun is the source of all energy. Light generates growth, unleashes movement, triggers ebb and flow. Clean Energy is solar power stored in hydrogen. But energy cannot be seen nor felt. It can only be experienced by its effect. In order to make this energy be felt by the visitors, we generated a wave surging through one of the old halls. The wave originates in the simulation of an energy force. A light impulse triggers the motion, which is then channelled by force fields around the hall's construction ribs. Finally, a dramatic ripcurl surges upward at the centre of the hall. The H2 theme is turns into a gigantic sensory experience - information becomes a living event.

The Bubble The wave has spilled out two drops, one of them splashing before the hall. This huge waterdrop residing in a pool called the bubble is the metaphor and supersign for the exhibition. The bubble as a cafe at daytime and a bar/club at night-time is creating a new focus for the public almost 24 hours a day. For the form generation of the bubble a special effect program was applied to construct the shape of a water droplet. The simulation generated the fundamental geometric form - the basis for all further derivations up to CAM production of the glassforms and the rips.

Digital Production Digital continuity from design to production is essential for our work. The spaceframe of the wave is derived directly from the threedimensional bend isoparms of the force generated surface. All beams are individually produced by CAM machines. A custom made aluminium profile is used to connect the joints and the membrane. The wave is the largest ever made spaceframe consisting of CAM produced profiles. 305 moulds were milled out of foam for the Bubble hull. Acrylic glas was bend over these moulds and cut at the edges using another CAM miller. The rips were jetstreamed out of aluminium. The bubble is a premiere of this construction worldwide.

干净的能源　设计展览馆所考虑的传递信息问题重于建筑，我们设计的并非装载信息的容器，而是沟通本身。BMW 集团 2000 年博览会方案从一开始就不是为了成就一个程序或机能，而是简略地沟通干净能源的概念；即指 BMW 集团所倡议在未来行驶时使用氢能源。氢技术由 BMW 完整开发，第一批汽车已经产生，但是氢气站与产生氢气的固定能源无法获取，因此 BMW 参与 2000 年博览会的动机，政治因素实高于技术问题。形式、内容以及表现全由我们去发展。

地址　经过 Verkehrszentrum 德意志博物馆比较之后，此展览馆被选定为2000年博览会的国际企划正式方案。地址选在市中心，曾是慕尼黑露天展览场。由于过去的混乱未被清除，此区目前犹如荒地，周围三个受到保护的旧回廊正在施工；回廊本身建于1908年，目前则处于非常糟糕的状态。由于我们的展览将持续五个月，我们决定着重于本方案的临时性。

波浪　活动主题"干净的能源—以太阳与水为驱动力"自己创造出它合适的形态。能量以不同状态存在，太阳是所有能量的来源，光引发成长、解放动作、触发退潮和流动；干净的能源是蕴含于氢气内的太阳能，然而能量不可视也无法感知，只能通过其影响来了解。为了让参观者感觉这个能量，我们设计了一个波，穿过其中一个古回廊；波来源于仿制能

AFFILIATION: ABB Architekten
COUNTRY: Germany / 德国

Internet Connection The Expo2000 itself is designed as a network of dislocal international events. The bubble is connected via the internet with the Cafe Del Mar in Ibiza, the favourite hangout place for the jeunesse doree for watching the sunset. A special style of music was developed there which will be brought to the bubble in realtime. Simultaneously the sunset of Ibiza will be projected on a wall in the West of the bubble. A virtual Ibiza sunset will superimpose the Munich sunset. Parties will happen at the same time on both sites with a shared experience and an two-way exchange of images and music. The event becomes dislocal. The information created form is site for an information based event.

↑ Computer-simulated model ／ 电脑模型图 → Interior perspective ／室内空间透视图

量，一道光触动运行，接着由回廊小梁的力场疏导，最后，一个戏剧性的回流，向上游至回廊中心；氢的议题转变成强烈的感官经验，信息成为活的事件。

泡球 波浪溢出两滴水，其一在回廊前溅开，这个属于水池的大水滴名为泡球，象征并隐喻展览本身；泡球白天为咖啡馆，夜间为酒吧／俱乐部。泡球形态的产生通过特效程序来制造水滴的形状，这项仿制引出基本的几何型态，继而成为以后发展、制造玻璃形态与回流的根基。

数码生产 从设计到生产，数码一直是必需品。波浪的空间桁架直接衍生出表面弯曲的 3-D isoparms，所有的梁皆分别由 CAM 机器制造，订制连接薄膜与接缝的铝制品；此波浪是有史以来由 CAM 制品组成的最大的空间桁架。305 个模具由泡沫材料加工成为泡球的壳，塑胶玻璃模具上方和边缘均使用 CAM 切除；回流自铝部位喷射涌出，这是全世界首次制造这样的泡球。

网络连接 2000 年博览会本身设计为分散的国际网络活动，泡球由网络连接至观赏夕阳的地方— Ibiza 的 Del Mar 咖啡馆。特殊的音乐即时传至泡球，同时夕阳将会投射到泡球西边的一道墙，虚拟的 Ibiza 夕阳将会与慕尼黑的夕阳重叠；两个地方同时举行派对，分享双方交错的影像与音乐。事件因而分散，信息引发的形体即是信息引发事件的场所。[翻译 王俐文]

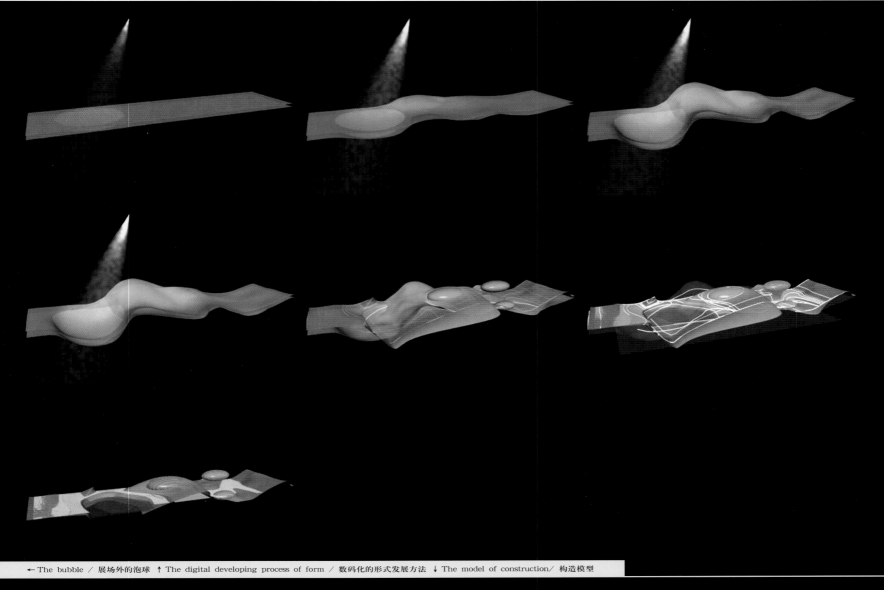

← The bubble ／ 展场外的泡球 ↑ The digital developing process of form ／ 数码化的形式发展方法 ↓ The model of construction／ 构造模型

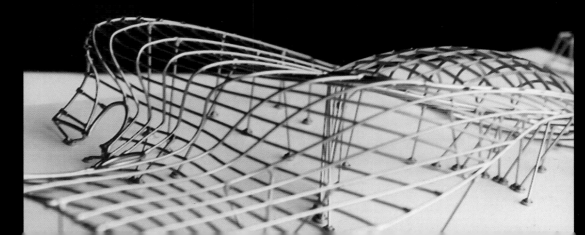

mediaThekmelbourne / 媒体中心

DESIGNER: Eugene T. H. Cheah

"mediaThekmelbourne" is a proposed centre to bring together the production and consumption of digital media. The facilities for the public exhibition of mediated art will include auditoria, external broadcast screens, media booths and immersive capsules. The building is conceived of as a series of pods clipped to an assemblage of three overlaid structural systems, sited on Melbourne CBD's Southern edge, alongside the Yarra, reclaiming a cut away piece of the river bank.

Digital Speculations   The architecture will become an interface for a distinct art form. It will seek to explore the implications of digital environments and digital processes upon architecture. The search is for a confluence between the digital realm and the material - a digital convergence. It will be a hub, to bring together the dispersed bits that make up the study, innovation and creation of digital media. The building is to be a visible node of the invisible network.

The internalised processes of this media will now be turned inside out, externalised to provide information and excitement to the public, the building variously serving as spectacle, interface and repository of knowledge.

New Topology   The building is sited on the turning basin of the Yarra river, pressed between the elevated train tracks and the river edge.

It is a void carved out of the riverbank in recent history, profoundly changing Melbourne's aerial edge profile. The physical site is a forgotten piece, thrown away and now rediscovered and reclaimed to redefine the edge for the CBD. The lost piece is draped over the built form as the new roof sheet, a soft plane shaped by the fluid flows of traffic around the site. A fluid dynamics study of the traffic pattern yielded the contour pattern, which was then interpreted into a composition from which to derive floor plates and form.

The topology is no longer concerned with the physical, it is concerned with invisible waves, of metaphysical information maps. Just as Paul Virilio's describes a disappearing city, where constructed geographical space has been replaced by chronological topographies, where (im)material electronic broadcast emissions decompose and eradicate a sense of place.

This notion of flows, implies a filter metaphor for the building, as a sieve on the Southern edge of the city. The filter striates the spaces, layering it parallel to the city and creating three rows of steel structures, each a self-contained ordering system with its own fractal rhythm and function.

Visually the object will add a glowing mass of screens to reconstruct a portion of the cityscape Southern elevation, currently dominated by expressway columns. Urban renewal for a site between elevated train tracks and the river calls for a programme that can thrive under such conditions. Spaces are immersive and discrete, each a pod or capsule that alienates from the surrounds. Immersive media booths surround the audience in virtuality and negate the noise, whilst the observatory allows views up and down the bend of the Yarra.

(Infra)Structure   The building accommodates public thoroughfares, two public-access auditoria, the larger seating 180 and the smaller seating 90. These can be combined into a continuous space. Digital media is presented through eight immersive environments and 33 mediated booths located along the walkways. A cafe and observatory is suspended out over the Yarra River. Finally eight smaller screens (one from each immersive environment), the south screen and roof shell provide mediated surfaces to the external public realm.

媒体中心意欲结合数码媒体的消费与生产。

数码思考   建筑会成为不同艺术形式的界面，探索数码环境及过程对建筑的影响。

新地形学   地址在雅拉河河谷，介于铁道与河岸间。

新的地形学概念，不再只囿于实质地形对建筑物的影响，而拓展至地形学上隐喻对建筑物的影响。

公共设施建筑物内有公共大道和两座分别容纳180人及90人的演讲厅。 8处室内空间及33座媒体亭均旨在传达讯息，咖啡座及观景台可远眺河谷。

数码观点   媒体中心的特殊结构及空间安排，会赋予我们日常生活更多新意。

建筑界面   通过虚拟实境来挑战对于建筑、空间及居住空间的已有观念，媒体中心实际上是虚拟及真实空间之间的界面，它包括展示间的展览厅，具有光电效果、常变换的玻璃屋顶，并面对河岸。[翻译 赵梦琳]

AFFILIATION: Faculty of Architecture Building and Planning, University of Melbourne
COUNTRY: Australia / 澳大利亚

Digital technology, forms a fundamental programmatic issue and its specialised spatial and infrastructural needs will be re-examined. Just as other services dictates have implicated upon built form, so too will the central server and fibre network be a deciding factor. The purpose will be to, paraphrasing Bernard Tschumi, construct a building at a time when the technology of construction has become less relevant than the construction of technology."

Programmatically the project encompasses all conventional activities related to the public display of art. Yet, the traditional notions of storage, exhibition and interaction are changed by a media that requires virtually no physical space to hold. The interaction between viewer and artwork is changed by the individualising character of immersive technologies. The building no longer serves as social condenser. The technology is one of disembodiment and abstracted communication - and hence abstracted community.

The building is ephemeral, serving as basic (infra)structure. The programmatic pods are clipped to the structural systems, configurable when necessary. There is no longer a need for expansive space as in the past. Nor for solidity, mass and enclosure.

Digital Narratives　The experiences that give shape and texture to our routines are being transformed by this medium, with its distinct structures and spatial arrangements.

Characteristics of digital environments, their experiential qualities and navigational systems, are explored as a source code for the making of architecture. Hypertext - non-sequential, non-linear narrative - is used as an organisational matrix, the hypertextual model of interconnected nodes informing the spatial composition. The building is conceived of as interconnected nodes of activity, discrete spaces linked by a series of walkways - circulation as a metaphor of hypertextuality.

The primary conditions of digital environments as immersive and interactive spaces provide the impetus for the resolution of the programmatic pods. These are essentially synthetic environments, for activities which are not spatially bound, network connections becoming as important to us as bodily locations, accessibility depending ever less on proximity.

Architectural Interface　The design seeks to challenge existing notions of architecture, space, and the inhabitation of spaces through the creation of the virtual. Yet the inhabitation of virtual cyberspace exclusively, divorced from physical reality, is limiting. Instead, the building is a physical container to hold an abstracted reality. There is, therefore, a search for a relevant type, an architecture that will adequately be the interface between the physical and the virtual.

Mediated booths provide the means of exhibiting the art. These visual prostheses are strewn throughout. There are two typologically different varieties of mediated booths set along the

walkways - one configurable, participative and enclosing; the other an external interface.

The roof shell is an ever-changing membrane of coated glass panels. These are opacified by electricity when necessary. At other times it is variously translucent, transparent or a projection screen upon which moving images are projected from the scaffolding structures.

Facing the Yarra River is the active face of the building, with screens that are remotely linked to the computers of syndicated artists, whose work in progress will be broadcast across the river, externalising the production of digital media, be it a blue screen of code and script or a rendering in progress. Strewn intermittently within this and protruding out over the river are the immersive environments, whose screens face the river and are visible from within and without, hence an internal/external dichotomy. Yet viewers within will be oblivious to the river they face, electronically extended bodies surrounded in virtuality.

↓ Site Context ／ 地址涵构

↑ External Perspective ／ 外观透视　→ Perspectival Sections ／ 剖透视

↑ North Elevation ／ 北立面图　↓ South Elevation ／南立面图

Martial — Art Headquarter / 武术总部
DESIGNER: William Hsien (project designer) + Koon-wai Wong (web page designer)

As a martial art practitioner for the last five years, teakwondo has impacted my personal life both physically and psychologically. Martial art has becomes a way of life : body, mind, and spirit are the three essences in practice. The ideal state in taekwondo is to achieve the balance between the three.

The initial idea of the building is to develop the project with the three essences in taekwondo. Take the body, mind, and spirit and try to develop them into a dialogue, so they can communicate with each other in the architectural language. Hoping they will achieve the balance as they do in martial art.

As the concept evolves, the digital tools allow the images of a kicking-motion to transform into a rough shape of the building envelope. Then the skin, core and structure are given according to 3D shaped of the building.

Skin, core and structure become the three essences for the building, and they define as three physical elements in architecture. The round glass element in the center of the building represents the mind. It is located in the center of the building as a metaphor that the center of the universe witch governs body and spirit. The curved aluminum panels on the exterior represents the muscle and skin of human body. It is flexible to express any motions, yet strong and hard to protect us against any harm from the outside world.

The building now has definition of human characters. It is almost like a living machine that defines through taekwondo philosophy and architectural language. It represents the spirit of martial art. And it will live as a martial artist who will always in search of the perfect balance in life.

习武五年当中，身心受跆拳道影响极深，武术已成为我生活的一部分。练武时身心魂必须合而为一，跆拳道的目的其实就在于平衡这三者。

我的灵感正源自跆拳道的身、心、魂，令三者交流对话而应用在建筑上，终于能达成一种平衡。

建筑物外壳源自跆拳道踢腿的动作，再据此发展出建筑物表层、核心与框架。

建筑物表层、核心与框架有如跆拳道的身、心、魂，建筑物中心的圆形玻璃象征心这个统治身、魂的宇宙中心，外墙的铝板有如人体的肌肉、皮肤；这既表达了肢体语言，也保护人体免于外部环境的伤害。

因此建筑物好像人体。它是一部活的机器，诠译了跆拳道哲学及建筑语汇，也代表了武术精神，犹如一名毕生追求着生命平衡的侠士。[翻译 赵梦琳]

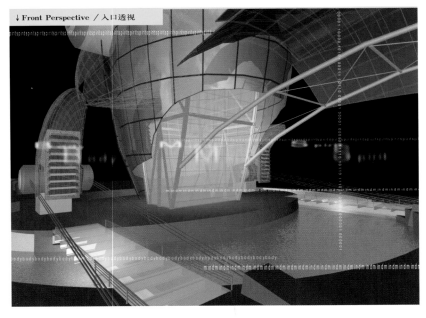

↓ Front Perspective / 入口透视

AFFILIATION: University of California, Berkeley
COUNTRY: U.S.A. / 美国

↑ Front Perspective ／ 入口透視

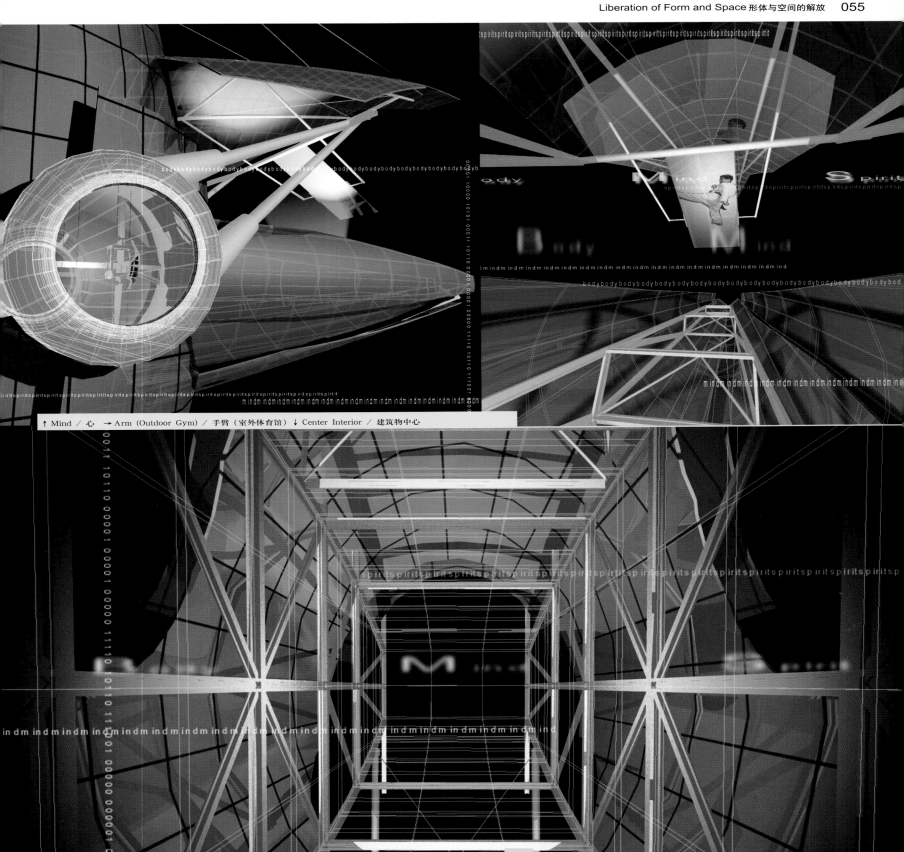

↑ Mind ／ 心　→ Arm (Outdoor Gym) ／ 手臂（室外体育馆）　↓ Center Interior ／ 建筑物中心

The Parasite Theater / 寄生剧院

DESIGNER: Shi-cheng Huang / 黄士诚

This project is to design an experimental theater for Chinese Comic Dialogue and its development and structure is basic on the original structure of Chung-San Bridge which located in Yuan-San, Taipei, Taiwan. The base is assumed a situation, which is the same as recent condition of Chung-San Bridge without considering laws and regulations or possible future changes, but discussing combining parasite concept with biology, architecture and performing art to create an experimental space design.

This project is beginning with the structure of Chung-San Bridge, the concept of parasite and the performance of Chinese Comic Dialogue, then debating the combination and transfer of elaborate contents and conducts of Parasite and Chinese Comic Dialogue. The audiences of this theater are from passengers of Chung-San Bridge or people who have activity in near by public space in Yuan-San, Taipei. The developing space of this theater depends on the original structure of Chung-San Bridge and with the interaction and metaphor of performance, pronoun, and intention of Chinese Comic Dialogue.

The main activities, not for long periodical, in this theater mostly to be held in the evening, and attract people mostly by rotating the theater in the beginning and in the end of the performance.

The whole moving designs, such as, stage structure, audience seats and stairs, are presenting a contrast to the parasite phenomena, seizing and replacing the host, and one of Chinese Comic Dialogue's tricks, "shaking bags". Shell shape virtual stage setting can attract audience by projecting images during changing scenes; pupa shape studio which located under the stage and embedded into the bridge can provide independence and concealing space. Before performance, the projection on the virtual stage setting would have attracting crowds around the area; at the beginning the performance, it would keep attract people by different moving spaces and changing shapes of the theater. In the beginning, the theater only just a guest to the bridge, by chance, it grew and developed by parasitizing; moved and expanded by using; acted as the host by performance.

Analysis of the words: [1]Biology An organism that grows, feeds, and is sheltered on or in a different organism while contributing nothing to the survival of its host. [2]One who lives off and flatters the rich; a sycophant. [3]A professional dinner guest, especially in ancient Greece.

ETYMOLOGY: Latin parastus, a person who lives by amusing the rich., from Greek parasitos, person who eats at someone else's table, parasite., para-, beside. See entry PARA-1., +sitos, grain, food.[1]

The parasite phenomena vs. architecture space　　According to the definition, in this project, "parasite" used as parasitic space lives in an existent architecture, takes it as its base, and gets tentative or permanent co-exist with the original architecture, the host, by expending, invading and conjunction. In the parasite phenomena of architecture, parasitic space must balancing co-exist with the host space.

此方案位于台北圆山中山桥，利用原有桥体来构筑相声剧院，而暂不考虑建筑法令或未来桥体变迁，但结合生物、建筑的表演艺术来创造一座实验剧场。

设计上利用移动的舞台，观众与楼梯来与所谓的"寄生剧院"形成对比，也用来象征相声艺术里特有的主客易位与"抖包袱"。在换布景时，会有影像投射至薄壳状的舞台，用来吸引观众视线。桥下舞台的空间为独立及隐蔽的。表演前投射到舞台的影像会吸引观众，直到表演开始，舞台均会继续移动，然后舞台会朝向中山桥，呈现寄生的动作并成为表演的主角。

所谓的寄生　　寄生：[1] 会生长进食的组织，藏匿于其寄主的组织中，且对寄主并无贡献。[2] 阿谀奉承者，依附他人而生。[3] 古希腊特有的专业食客。

拉丁语词源：靠富人生存的人。　parasitos 原指古希腊富人餐桌旁的食客。"para 专指"一旁"，"sitos"指"食物"。

寄生现象与建筑空间　　此方案的寄生指依附于现有建筑所形成的空间，与现有建筑共生，连接甚至侵入现有建筑。而寄生建筑的空间需与寄主的建筑空间平衡。[翻译 赵梦琳]

Notes / 附注
[1]The American HeritageÆ Dictionary of the English Language, Third Edition. Copyright © 1996, 1992 by Houghton Mifflin Company. Published by the Houghton Mifflin Company. All rights reserved. http://www.bartleby.com/61/

AFFILIATION: Institude of Architecture, NCTU / 中国台湾交通大学建筑研究所

↓ Physical model (closing) / 模型（闭合时）

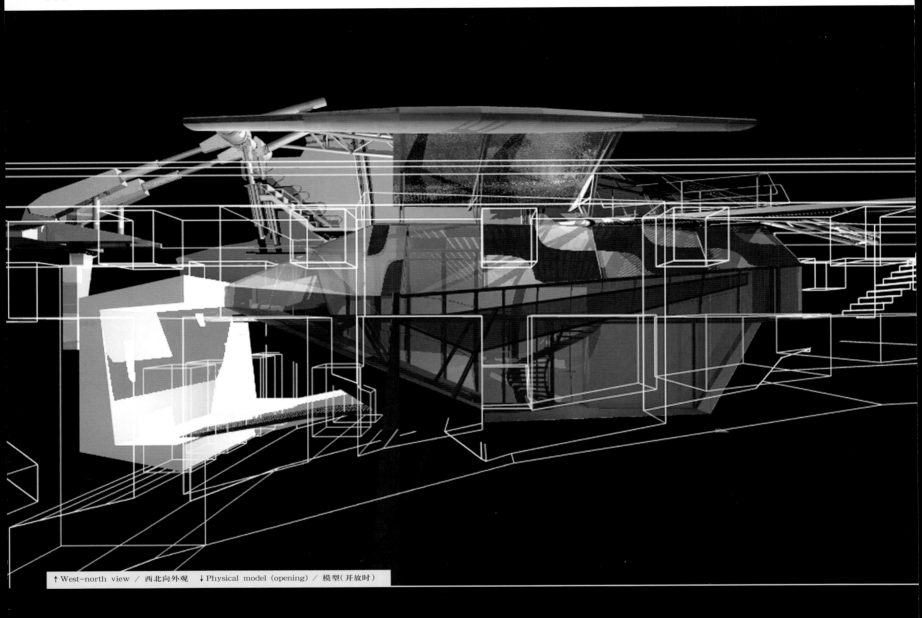

↑ West-north view ／ 西北向外观   ↓ Physical model (opening) ／ 模型（开放时）

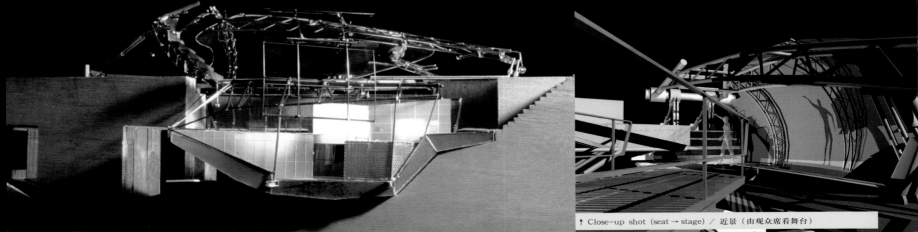

↑ Close-up shot (seat → stage) ／ 近景（由观众席看舞台）

Idea "Cloud" —
An Experimental Dance Theatre, Grange Beach, Australia / 云
DESIGNER: Tristan d'Estree Sterk

Idea "Cloud"　This project is about a whim. About building a cloud; building something that shouldn't and can't be built. It's about escaping architecture and about producing a space that can only be truly understood as a happenstance of circumstance, an instant, a moment in time. It is also a project about a new way of thinking. Architecture as an envelope. Architecture as a circumstance. Architecture as an order that emerges from complexity. Architecture as field. Architecture that is of a non-geometrical order, that is ephemeral. A whimsical order. As an experimental dance theatre, it is a project that begs the question, in what new ways can our bodies engage with spaces that are influenced by the contemporary condition 1?

[1] The contemporary condition is one influenced by the prolific use of information technologies. Several authors suggest that the contemporary state is one that suffers from dislocation, navigational anxiety, and the new market commodity of thoughts. It is an ephemeral society, measured and understood in terms of traffic flux, conditions of acceleration, and field theory. It is a society that is learning to challenge the object, and cherish relational databases.

The Condition of Information (In Brief)　The root to understanding the synthetic comes with the realisation that information has a construct that is based upon a system of relative relationships between sometimes disparate pieces or bits of information. We witness the imposition of relative information orders onto space constantly. For example, think of how you arrange your office desk, think of how you order your diary, your hard disk, your note pad, your office. All of these things are divided into parts with different places for different types of information. In such systems, tree-like frameworks emerge as natural phenomena in part because information exists within a system of relative ordering, which although spatial, don't use space in the way that architectural ordering systems or "object" based orders do. This is because information loses its validity or meaning once it is removed from a relative environment, an environment where one piece of information can be weighted against something else. This obviously isn't the case with a geometrical or architectural order because object forms have the ability to stand within their own right as objects of beauty. Information simply can't do that. This very simple example opens the door to understanding the current information environment and some of the cultural, spatial and architectural implications of information systems.

Information Systems as Fields　Information Technology systems, introduce a new, object challenging philosophy into society. Such systems don't rely upon the object, but relationships between objects or bits of information. In these systems, things are always relative to one another and sometimes expressed as mediated occurrences 2. Things within these systems never stand upon their own, but rather within an intricate system of relationships, within a sense of plurality, ephemerality, and total complexity. Information societies are based within notions of field. This is why built environments are increasingly being measured in terms of traffic flows and ever increasing mobility. TV, telecommunications, air-traffic transportation and computer networks have altered our spatial perceptions. The architecture of fields is ideally suited to study contemporary complexity.

[2] Mediated occurrences within information systems usually are associated with experiences of relationships between data that are more ephemeral. Temporary bits of data, event data. The news broadcast, the weather report, the video-cam view of a distant city, or web-tv images of the gulf war are all examples of mediated bits within information systems.

Clouds and Fields　The metaphor of a cloud has been chosen to propel this project. A cloud is a field. A field is a complicated, whimsical, ephemeral, system of balances. It is an envelope or zone in which the possibility of "cloud" emerges. Clouds are a condition that can be understood as an object; But a false object at that, because really, it is a system based upon a series of localised relationships between particles of dust and water. It is a happenstance of circumstance that we can recognise and associate with a temporary structure that changes as it blows in the wind.

AFFILIATION: The Office For Robotic Architectural Media
COUNTRY: Australia + Canada + Holland / 澳大利亚 + 加拿大 + 荷兰

此设计是灵感的体现。我们说要建造云彩其实是要盖不可能的建筑，设计一个偶然的空间来逃避建筑。

信息和空间不同，信息其实与其环境非常相关。建筑与物体可单独存在，但信息不行，它们二者之间的关系。

信息领域（场所）
会很大的挑战，它的重点非物体本身，而是信息与信息间的关系，系统中信息互相依存，并常以思考型事件出现。

云与场所（领域）　选择"云"做为隐喻。云即为场所。云既复杂、灵巧、智慧又处于平衡，云是物体也是虚幻的，是一系列灰尘与水汽的组合。我们因此采用会因风改变形状的临时建筑。

空中之舞　云在任何形况下都会成形。它们轻巧，会以任何形状出现，在室中旋转、飘浮、聚集，任意蒸发、成形、趋近水平线。云朵彷佛在做实验性的舞蹈。

场所　云朵的观念与场所的环境也息息相关。[翻译　赵梦琳]

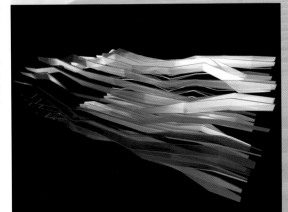

↑ An emergent structure taking shape. / 正在成形的结构

Fields can be thought of in terms of clouds. They are spaces in which orders arise from a balance of forces to produce a recognisable pattern. Of course field theories and the conditions that they bring to form making, are all motivated from the bottom-up, they are circumstantial to calculations, with no creative act required to define the architectural form. And in that respect there are no overarching geometrical notions or understandings that can be used to come to grips with exactly what the space is about. The spaces can only be understood as a circumstance, or a condition in which something may occur. As within the post-minimalist art movement, the architecture of a field challenges the notion of what is important about the act of creation. It questions the ambition of geometrically ordered form and suggests that the art of form making rests within finding the balance in which conditions of relative stability occur, and it is the challenge of the architect to channel and mould forms within that occurrence.

The cloud won't occur outside of the parameters it requires for its survival. When it is within these parameters it can survive, but is moulded and blown by the wind, into an ephemeral, ever changing structure. Hence the emphasis exists not so much within the cloud's form, but rather the notion of how the form has come to exist as a balance of circumstance.

Dancing Across the Sky   They come to being out of environmental conditions at any time, in any place wherever the circumstance provides. They are whimsical, pushing at the limits of their own existence, spinning and weaving and soaring and storming through space. Evaporating; they challenge the boundary of their envelope, pushing perhaps towards the line of an intangible horizon. Clouds are an ideal vessel for experimental dance to occur within. Looking at the way in which the motions and forms of clouds are driven by field conditions, it only stands to reason

that the spaces in which performances occur should also be driven by field conditions. The spaces should be like a sky through which the conditions of "dance" emerge. Dance is ephemeral. Dance comes to being at any time, in any place wherever the circumstance provides.

Field Performance   It is at this point that the field conditions inspired by notions of "cloud" touch upon the ideas associated with fields of information. Both fields, that of the information system and the natural form generating system of clouds overlap to produce a rich and challenging space.

Dancing within and through fields exposes experimental dancers to a creative act that is instinctive and ephemeral. Things aren't constant within fields, but rather a shifting balance of free flowing opportunity. Things are spontaneous, and as a performance comes to being, the dancer has the option to move through the space, opening new possibilities for the development of their work.

The performance fields within the space are defined by the three complimentary information technology nodes, a lighting node, an acoustic node, and a projection node, each of which have associated fields. The nodes are arranged in a fixed position within the cloud, but their individual fields can vary according to the dancers' choice of perimeters dedicated to each field.

Each of the three fields overlap as to produce either a tightly knit or sparse almost non-existent envelope in which the performance may occur. Dancers engage with the fields incorporating elements of projection, lighting or sound into their work as they see fit.

Around each performance space, people gather to watch the event take place. The spectators themselves relate to the field of the dance to produce an envelope of spectacle. Yet, because fields change, because they are ephemeral, because they are a balance of circumstance, the audience may unwittingly be challenged by the instinct of the dancer and become a part of from field in which the dance seeks it's inspiration. Audience becomes dance. Everything is spontaneous, forming just as a cloud does within a zone of possibility. The forms that bodies take, become whimsical and dance becomes understood a happening that emerges from an envelope of circumstance.

The Site   The beach, the iconic Australian site is a field. It is a place of condition, influenced by the complexities of hydrodynamics. The construction of the beach, its sands and contours is temporary; it is a place of movement and is built according to the constructive and destructive actions of waves. Beaches emerge from the delicate balancing act between environmental conditions, tidal and seasonal variations and the actions of man. It is a place in which people enjoy and appreciate the effect field conditions have upon the Australian landscape. And above these shifting sands, hovering over the delicately balanced landscape, is the sky, dappled with clouds.

↑ Exploded diagram of theatre space showing services (red=a.c. yellow=IT/electric). / 剧院内服务空间的示意图

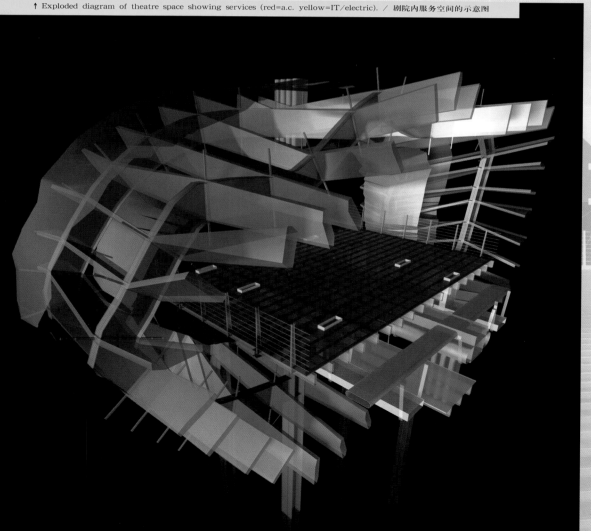

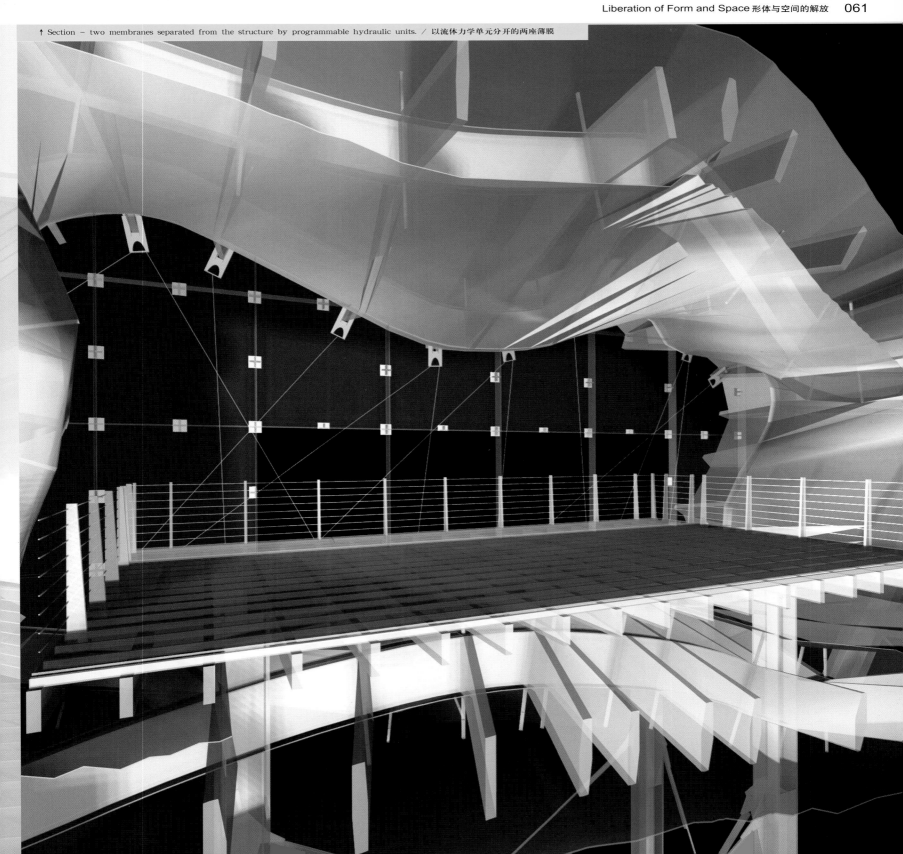

↑ Section – two membranes separated from the structure by programmable hydraulic units. ／ 以流体力学单元分开的两座薄膜

Architecture in Transition — Energy Container / 过渡中的建筑—能量容器

DESIGNER: Hsin-hsien Chiu / 邱信贤

The Project is composed with 3 processes:

Process 1: Operation

In the beginning, we take the sound wave graphic records to trace the whole sound energy transformation in the same period by media-model study. At the same time, by observing the interaction happened in different material / medium, we want to discuss the interaction and transition between energy and space.

Process 2: Site Impact

Finding out some place in the city interacted with hybrid energy-It used to be the residential area before. Because of the viaduct, which was high-press inserted, the back parts of the house become the most important facade to the road. Besides, the character of the space is transformed from pure residential area to commercial one. On the other hand, the people and the cars are forced to impact here for the extension of heavy traffic.

Process 3: Urban Energy Container

The design tactic is to extend and connect the activity of human by inserting an urban park higher than the viaduct. Besides, we inject some part of the traffic on the viaduct into the energy container, and the people and car from the viaduct will go for relax and get some leisure time inside the container. Whole the concept above are concerned to become the program of the design. Hopefully, after the entrance of every energy interacted here, it can proceeds for some energy transition and the energy will be unleashed after totally renew.

Foreword *"Energy / Which exists in different kinds of status / When materials start to interaction / Energy will begin to transform to another mutation*

2001 / The interaction of the centuries / Everything will face to a whole new life / Architecture / May have difference with the past / May not be so clearly anymore / But / In a fuzzy zone / Lurk some mutation / transition" — Architecture in Transition

本设计方案以三个步骤创作：

步骤1：操作

我们最初运用媒体模型，研究声波的图表记录，以描绘整个声强能量在同一时期里的转变。同时观察不同媒体或媒介之间的相互作用，试图探讨能量与空间之间的过渡与相互作用。

步骤2：场地冲击

我们找到都市中某个与混合能量相互作用的场所。它曾经是住宅区，高强度建造的高架桥使得建筑的背立面成为面向道路最重要的面。此外，本区也由纯粹住宅特性转变为商业性质；另一方面，由于车流而造成人潮与交通的冲击。

步骤3：都市能量容器

设计的策略是置入一个高于高架道路的都市公园，以延伸并连接人的活动。我们也将高架道路一部分的交通加入此都市能量容器中，让人可在此休息片刻。以上所考虑的事项将会纳入整体设计，以期加入的每一个能量在此相互作用后，能造成某些过渡，在完全更新后，能量得以释放。

序 "能量 ／ 以不同的形式存于 ／ 当物质开始互相影响时 ／ 能量开始转化突变

2001／ 世纪交替 ／ 一切面临新生／ 建筑／ 也许与过去不同／ 也许不清晰

但在混沌之中／ 正在转变与过渡" —引自《Architecture in transition》[翻译 蔡咏岚]

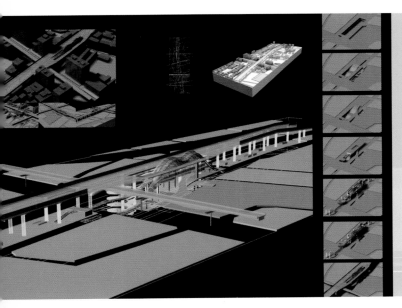

AFFILIATION: TamKang University Architecture Graduate School / 中国台湾淡江大学建筑研究所

Imaginary Building(So Far So Close) / 忽远忽近的想像建筑

DESIGNER: Gianluca Milesi

This project is about an imaginary building. This is not linked to a particular location or to specific environmental conditions.

The genesis and the conception of the idea is strictly connected with a computer elaboration; Curved lines extruded in the space generate complex enclosed spaces; these spaces intersect themselves in several points. Outside and inside unusual geometries drive the users through an imaginary, hopefully poetic, but real condition of perception.

As a building the structure is thought as a series of skin membranes modulated to a small scale in order to reach a large variety of solutions by an addition of simple and repeated elements.

The project is a sort of exaggeration and aberration of geometry; the design is an experiment in between fantasy and a real process of construction and is a step part of a larger research focused on a non traditional and alternative way of building, more suitable to non regular and non monotone conceptions, to more flexible constructions.

The idea is to introduce concepts or attitudes of thinking architecture as "membrane", "skeleton", "inside-outside", "organism"(not in the natural organic sense), "self-sufficient surfaces", "high and low technology", displacement, time and space perception, process.

This is an attempt to escape the rigidity of the generally meant serial architecture and the authority of traditional and 'rational' technologies.

As the last issue is interesting to point out that an approach as this is apparently detached from a real world or a real city creates new possible (even if contrasting) relationships the urban environment, suggesting the new 'imaginary scenery' of the city.

Cities now probably need this new contrast to pass to a new condition, enriched, representing the new spatial and visual paradigma.

该设计其实是一个想像建筑。它实际不属于任何特定地点或环境。

设计理念主要源自电脑；曲线界定了复杂的空间，这些空间在许多处纠结交错；空间里不寻常的几何，让观者能感受到其诗意。

建筑物的结构是一系列的薄膜，但尺度较小，通过较简单而重复的单位来实现不同的解决方案。

该设计的几何夸张，失序，介于现实与想像之间，为了创造非传统的建筑，对非常规、多样化且具弹性的建造进行了实践。

该设计将建筑视为一层薄膜，框架、内外、组织、自足性的表面、高低并蓄的技术，位移，对时空的感知及一种过程。破除建筑理所当然的序列性及理性。

这样的设计方法脱离现实，却为城市及都会环境提供新的关系与情境，现代的城市也许需要这些刺激，丰富并提供新的空间及视觉典范。[翻译 赵梦琳]

↓ Interior views / 室内

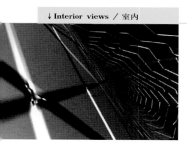

AFFILIATION: science-architecture, New York
COUNTRY: U.S.A. / 美国

Cinematic Architecture / 影像建筑

DESIGNER: Tom Munz

This investigation of Film and Architecture is looking at the quality of an experience in a space. One often watches a film and is stimulated emotionally. Our emotions move up and down, from laughter to sorrow, excitement to fear. Although this has to do with the connection we have with the characters, even stronger is our visual connection with the rhythm, movement, lighting, colors, sound and composition in the film.

The main issues attacked are, if elements in film can be reinterpreted into architectural space, while maintaining the experiential qualities presented in the film and is it possible to create a narrative through architecture? By deconstructing film in order to better understand its two dimensional elements, is it possible to then reconstruct these elements into a three dimensional space? Someone that visits a piece of architecture has an amazing amount of freedom in the way he explores the architecture. This is opposite to film where you as the viewer surrender control giving another permission to manipulate you. Thus, by looking at this new media there will be an understanding gained about space, which may not have been seen otherwise.

This final analysis examines both how a film inspires new ways of thinking and how it can manipulate us. Here the design relates back to a scene from "Matrix". Efficency in use of space and using technology to conserve the environment are ideals explored in the design. A connection is also made to becoming part of the technology, plugging into the system, sharing information and ideas. Finally, how film takes away our control. We must surrender control over what we see and hear when we experience film, as in the constructs of this Internet cafÈ, one must surrender his ability to go where he wants as he is hoisted into the air. This is just part of what can be taken from film to make dynamic architecture.

对影像建筑的探索是为了细究空间经验的本质。电影常激起观众高低起伏的情绪，使得悲喜惊惧交织于与演员甚至影片的韵律、光线与色彩的共鸣中。

建筑是否同样能引起观者的共鸣？解构电影的二向度元素，是否能在三维空间中重新组构起来？建筑中的观者能够自在地漫游，电影的观众却被时间和事件捉弄，也因此新的媒介将能透露出空间的讯息。

最终的分析验明了电影如何激荡着新思路，又如何操纵着我们。这个设计可追溯至《黑客帝国》的一个场景。设计的理想在于提高空间的使用效能及运用科技以保存环境。沟通为成为科技的一部分、与系统连接、共享信息而存在。电影运用视觉与听觉夺去了我们的控制权，就如同在这个网络咖啡店建筑中，观者在升向空中时必须先放弃想往哪里去的自由。而这只是能用于动态建筑的部分电影技巧。[翻译 赵梦琳]

↓ Two worlds come together / 两个世界合而为一

AFFILIATION: Ball State University
COUNTRY: U.S.A. / 美国

Differentiations and Repetitions — Housing in a Big Scale /
差异与重复—大型集合住宅

DESIGNER: Noboru Ota

The aim of this project is to investigate the new method to repeat and diffrentiate the housing system in a big scale.The project looks for the condition that several systems of housing, for example units,structures, public facilities, and circulations are interrelated and are transformed through the interaction of one system with another. Three kinds of deffinitions of housing system; Intermediate Space, Field of Equilibrium, and Structural Discontinuity make a feedback cycle in the design process.

Intermediate Space　The project looks housing units as connecting devise for repetitions,and sets up intermediate space related to spatial continuity between inside and outside. In other words, intermediate space is potential space which can be stateed as both inclution and exclution.

Field of Equilibrium　The project uses the topological condition of field for the transformation of repetitions.

Structural Discontinuity　The project utilizes the discontinuity as genesis of space which emerges with structure.

本方案探索了大尺度集合住宅中的差异与重复。审视集合住宅的几项元素：单元、结构、公共设施、动线及这些系统间的关系，我们将集合住宅定义为三种：中介空间、平衡场所及结构的非连续性，它们均为设计过程中回馈系统的一部分。

中介空间
装置，而以中介空间来连接内外，即可被纳入亦可被排除。

平衡场所　以拓扑学的模式作重复的转化。

结构的非连续性　以结构的非连续性做为空间的原型。[翻译 赵梦琳]

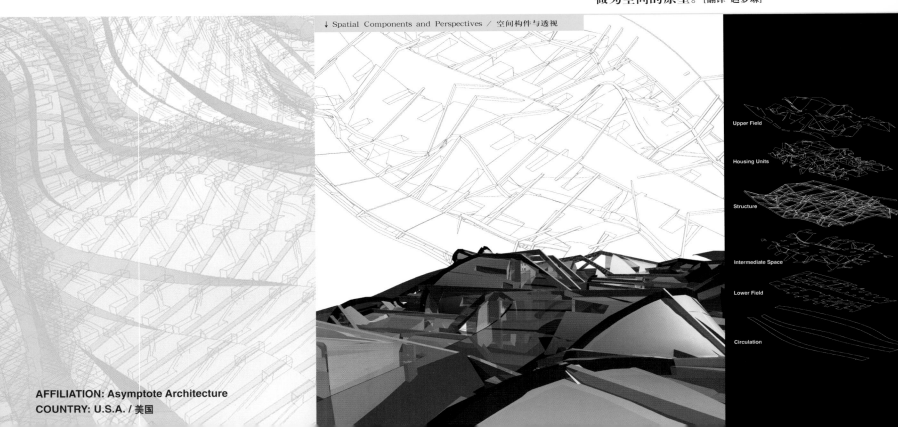

↓ Spatial Components and Perspectives ／ 空间构件与透视

Upper Field

Housing Units

Structure

Intermediate Space

Lower Field

Circulation

AFFILIATION: Asymptote Architecture
COUNTRY: U.S.A. / 美国

New Housing for New Bodies / 新时代的住宅

DESIGNER: Jesse J. Seppi + Vivian Rosenthal

The speed of the highway. Of endless movement. Of endlessness. Constant shifting of space between cars. An always almost violent dance. The threat of the crash. Of something new. Of permutation. The repetition fails. The paradigm of the car crash and the ensuing transformation structured this housing project.

The modules-based on the idea of the car as module-followed a linear movement from south to north: mimicing the movement of the nearby highway. As the modules move north they begin to crash into one another as they attempt to avoid existing debris. This spatial collapse yields a hybridized form at the north end of the site. An interstitial or residual space between these mutated forms is reappropriated as circulation. There is a simultaneous overlapping of bodies, flows of information and spatial forms. It is this simultaneity, this interaction between space and the body that produces digital space. This digital space renders the subject as object and the object as subject in that the dichotomy between the object (architecture) and the subject (body) is rendered obsolete. This digital space (which we are defining as interactivity; the space and the body act on one another and in turn re-act to one another) is one that is shifting, mutating and is always (becoming) something else.

高速公路上的速度、不停的运动、车辆间变换的距离，就像激烈的舞蹈，在新旧变换中，重复被打破，撞车后的汽车变形体成为本住宅设计中的结构范例。

此方案主要取材于车辆的移动及碰撞，基于模型及由南到北的线性移动，模拟邻近的高速公路现状。这些单位往北移动，但为了绕过废墟而碰撞。空间的崩溃在这一地区北端产生混种造型；这些突变造型内的残余空间与缝隙其实就是动线。造型内有同时交叠的量体、信息及空间，也就是同时性、空间及量体的交叠方能产生数码空间；这些空间可使主客物体易位。数码空间(我们将之定义为互动，空间与实体的互动)会变换、突变及变为其他形态的空间。[翻译 赵梦琳]

↓ Module Massing / 模矩量体

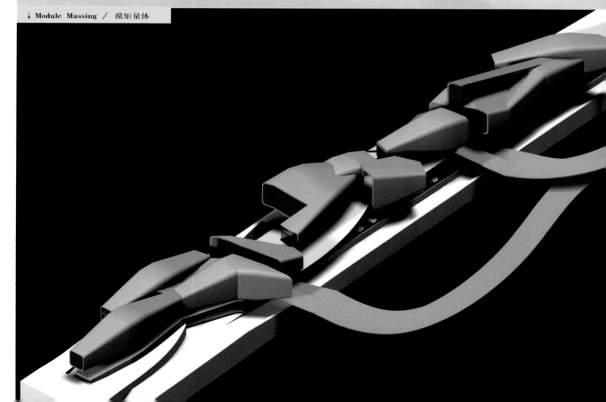

AFFILIATION: Tronic Studio (www.tronicstudio.com)
COUNTRY: U.S.A. / 美国

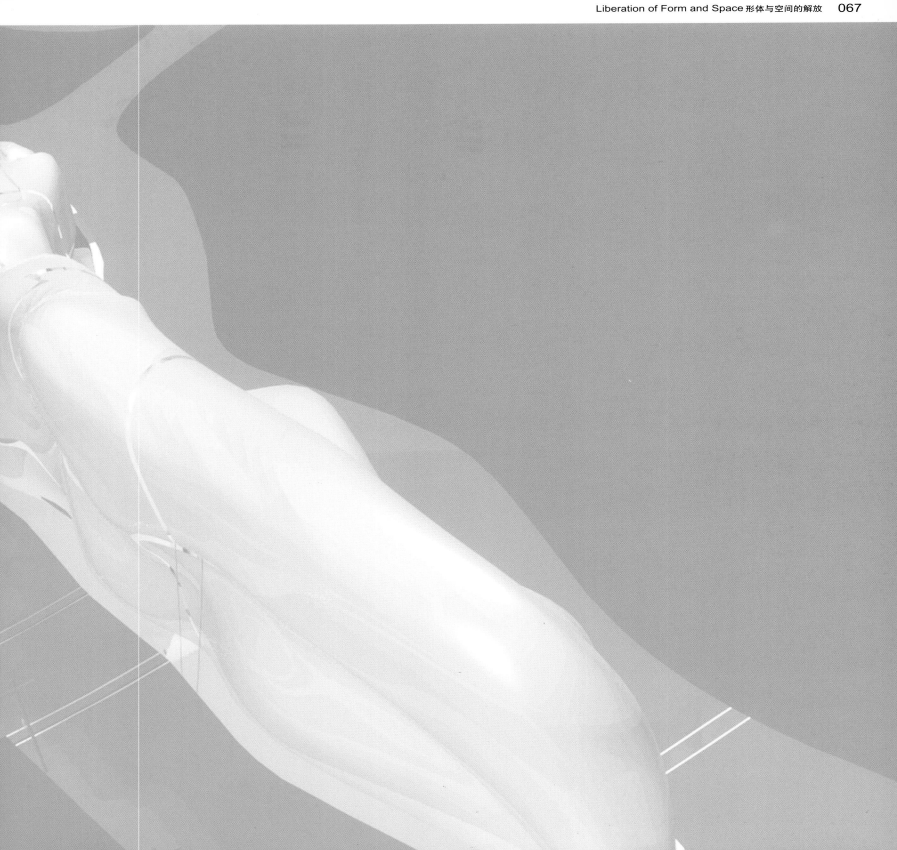

Parametric

Parametric Intelligence of Design

Frank Gehry once said that computers cannot create the curves of a deign, and are only a tool to help us draw the curves as they are in our mind. This is a good was of explaining what the role of computers is at present. But with computer technology developing so fast, who knows what they will be able to do for architecture by the year 2050. Will they be able to design their own buildings? The question recalls Allen Turning's 1956 question: "Can computers think? " People were skeptical then, but Deep Blue's artificial intelligence was able to beat the world's greatest chess player, seemingly answering Turning's question. In the future development of digital architecture, it remains to be seemed whether computers will have the ability to design structures themselves. In the near-future, computers will be able to handle basic decision-making when given a set of limited parameters. In the future, though, it may be possible to develop artificial intelligence and eventually a " digital intelligent architect ". From this perspective, we are only in the very early stages of digital architecture.

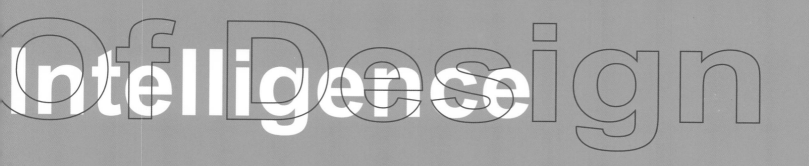

Intelligence Of Design

参数化智能的设计　　　Frank Gehry曾指出，电脑并不曾创造那些曲线，而仅仅是个帮我们画出那些线条的工具而已。这段话十分贴切地叙述了当今电脑对建筑所能扮演的角色。那么，以电脑发展的快速来推测，2030年或2050年甚至2100年时电脑能为建筑做什么？电脑能自行做设计吗？这些问题让我们想起1956年电脑刚刚诞生时，科学家Allen Turing所问的："电脑能思考吗？"。当时有人认为一定能，而有人却嗤之以鼻，但后来的"深蓝"人工智能程式，具有可怕的思考能力与知识，打败了世界国际象棋冠军棋手，从而使得1956年以来的问题不再是问题。因此，在数码建筑的未来发展上，对于电脑能否具有设计知识与设计思考能力这个问题，且让我们拭目以待。在近期，在给定条件的情况下，电脑将可以做出基本的判断。在将来，电脑可能在人工智能发展的基础上，成为一名"数码智能建筑设计师"。从这个角度来说，我们现在的数码建筑还是刚刚起步而已。

Digital Condominium / 数码公寓

DESIGNER: A.Scott Howe

Concept & Description   The majority of discussion regarding digital architecture assumes the computer to be a tool that can be used in the design process. The outcome is still considered to be drawings, images, and even animations of a building that exists in digital form. The "Digital Condominium" project takes the theory of digital architecture to new heights. The following declaration is made: *A digital building should know how to output a real-world counterpart of itself.*

Proposed is a "Digital Condominium" that is designed in cyberspace, uses online. factories to manufacture its kit-of-parts, is constructed with robotic building systems, and is managed during occupancy by its own virtual self. The condominium project consists of two major systems: the skeletal support superstructure and the plug-in infill module system. In this way the support structure becomes an infrastructure much similar to roads and utilities, waiting for infill construction to begin. Contrary to road systems, the support superstructure is fully demountable and recyclable.

Unit Plans   The individual residential units that would be inserted into the condominium super structure are not as easy to define. Since each unit would grow and constantly be redesigned by the owners and tenants, the architect has no business producing a static design that cannot change. Instead, the digital architect should design systems with rules and grammars. Design the system and let it behave as it likes without interference. The outcome will be more beautiful that way.

Digital Manufacturing   A compact conceptual factory has been conceived for manufacturing the modules. Kit-of-parts components are manufactured in anonymity on the Internet. Orders based on supply and demand are made by online agents. Prices rise and fall depending on free market bidding. A supply of certain components are delivered and queued for assembly. The compact robotic assembly line takes from the queue and assembles little by little the modules as they are ordered. Real time manufacturing. Conceivably, the modules could be transported by robotic trucks to the construction site. The factories could conceivably be digitally retooled remotely over the Internet, so location would be irrelevant. A homeowner would simply design the residence using virtual tools, and execute the manufacture of the components.

Digital Assembly   The Digital Condominium was conceived to be a demonstration project which showcases a new construction concept that calls for the establishment of a site factory. In the site factory concept, traditional "final line" construction is replaced by many parallel manufacturing and assembly processes in the "assembly line" method. This includes the assembly of building components on the site using robots and automated construction equipment.

Digital Facility Management   It is conceived that each dwelling would have its own independent server connected to the Internet. A facility-wide server would hold the entire virtual building with simple bounding-box representations of each apartment. Using the World Wide Web, modules would be downloaded from their source in VRML format, and assembled in real time on a visitor's browser each time the building is viewed over the Internet. Clicking on each of the dwellings would follow the link to the virtual building held within the server of that dwelling, and would show a VRML model of it on the visitor's browser, in the form of a facility home web page.

概念及说明   大多数关于数码建筑的讨论都将电脑假设为设计过程可使用的工具，其结果就是把数码化的绘图、图像以及建筑推上动画；"数码公寓"这一方案将数码建筑达到新的高峰，并声明如下：一个数码建筑物必须知晓如何输出自己在现实生活中的样本。

在虚拟空间内设计、使用的"数码公寓"。生产其配套元件的工厂由自动的建筑系统建造并在居住期间自行管理。公寓由两个主要系统组成：上层构造的结构以及填实模系统。

如此一来，支撑结构更类似于道路与公共设施，等着开始建造时被填满；与道路系统相反的是，上层构造可全数拆除并回收。

单元平面   位于公寓上层结构的个别居住单元不易界定，由于每个单元将会不定时地被屋主重新设计，建筑师没必要生产一个无法改变的静态空间；相反地，数码建筑师应该设计含有规则章法的系统，设计它，使它可畅通无阻地随意表现；如此以来结果将会更完美。

数码生产   构想一个小巧、概念形的工厂生产模具，为网络制造配套元件，其订单以供货量为基准，而需求由在线代理人设置，价格的涨跌由市场自由竞标。特定元素的供给品送达并排列好等候装配，小型自动组装线从中取得供给品后，一点一点地依其顺序组合成模具，即时生产，模具经由自动手推车转运至工地。

AFFILIATION: Assistant Professor (visiting) , University of Oregon, Department of Architecture.
COUNTRY: U.S.A / 美国

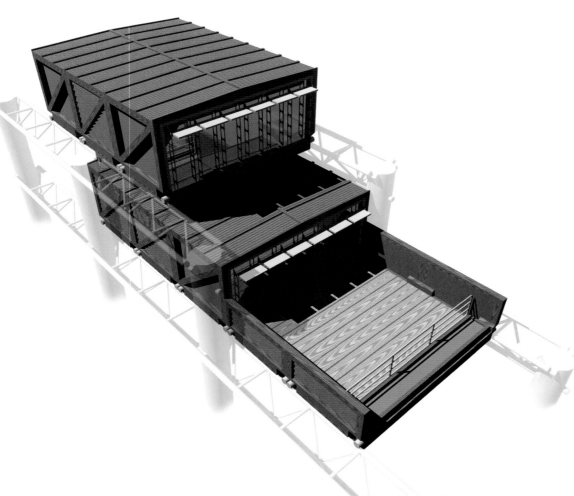

工厂可在网络上更换数码机械设备，因此地点灵活，屋主可以简易地使用虚拟工具设计居处，并完成零件生产。

数码配装　数码公寓成为展现工厂的新施工观念的示范方案。在基地工厂的概念中，传统的"最终线"施工由许多同时进行的生产与组装过程所取代，被称为"组装线"的方法。包括在基地使用机械与自动施工装备组装建筑物元件。

数码设施管理　每个居住单元有独立的主机连接着网络，全域主机简洁地表示了单元每个公寓的盒子，涵括整栋虚拟建筑物。全球网络使建筑物在每次被浏览时，可从 VRML 下载模具，即时在参观者的浏览器上组装；每点击上一个居住单元，便可连接到那个单元主机的虚拟建筑物，并以社区首页的形式在参观者的浏览器上显现它的 VRML 模型。[翻译　王俐文]

↑ Collapsible steel module ／ 可折叠的金属模件
↓ Plug-in condominium perspective view ／ 插入式公寓的透视

→ Interface with superstructure system standard on each module ／ 隐含标准特级结构系统的表面
↓ Conceptual factory for module manufacture ／ 制造模件的概念工厂

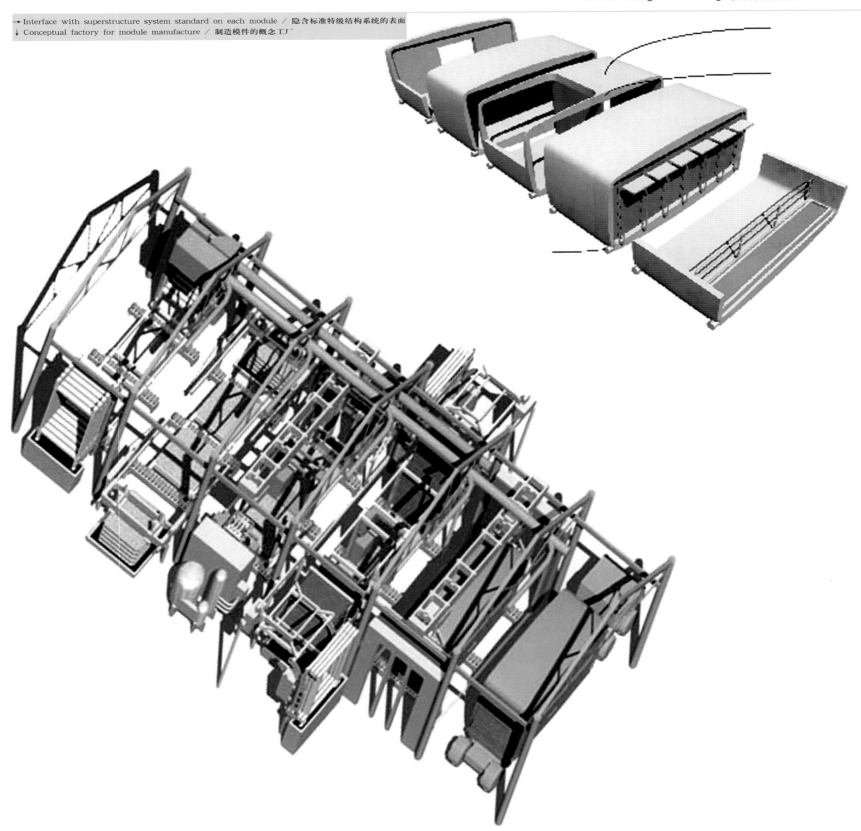

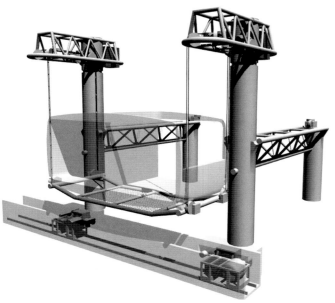

← Robotic construction system ／ 机械化的建筑系统
↓ Modules are lifted into place and plugged into the support structure ／ 模件被置入适当的位置，并插入承力结构体上

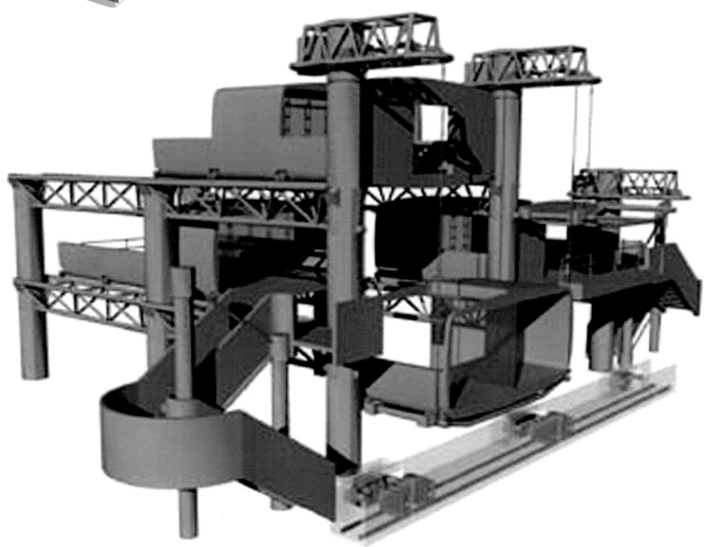

V-Mall / V 字购物中心

DESIGNER: Nonlinear Architecture, New York, USA
(Principals: Philippe Luc Barman, Gabriela Barman-Kramer)

The development of the commercial mall in the United States during the past fifty years has shown numerous variations of two main typologies: the 'main street' and the 'super block'. While the main street malls have proven to be more successful in the western part of the United States, above all in California, the super block conquered the east. The differences between the two are mainly based on cultural and climatic influences:

Main street malls simulate the small scale of American suburbia, providing an open public space with all the clichÈs necessary: street lanterns, rose bushes and plazas. This scenery provides an opportunity to downplay the purely commercial function of the mall and allows the visitors a break from reality.

The super blocks intentions are similar: a postmodern fortress of commerce. Like main street malls, they are placed safely in the middle of a parking lot buffer zone. Once inside these malls, each visual connection to the outside world is eliminated to get the visitor's attention for what is really important, namely to spend money. Even though both typologies are based on cohesion through artificial public space, they deny their public context.

The v-mall project for Los Angeles intends to merge commercial and public spaces into a complex that reflects a strong relationship with its urban environment. It can be accessed by foot from every street enclosing the site and through an underground connection from the Disney concert hall. An urban path winds through the v-mall, connecting internal spaces as well as the key urban locations outside the mall. Cinemas and parking are located on the lower levels while the retail spaces and the main public spaces occupy the upper levels where they are provided with sufficient natural light and visual connections with the city. The open public spaces become an urban condenser with a program defined more by the self-organizing forces of social behavior than by its architecture. These spaces can be used for outdoor cafes, performances, skateboard contests, flea markets etc.

The v-mall becomes an attractor for downtown Los Angeles together with the near-by concert halls and museums. Shopping, therefore the pleasure of spending money, happens in a more conscious way and on the same level as the public attractions that are free of charge. The retailers profit from the urban condenser and, in return, make the v-mall profitable. The residential area and the urban institutions next to the site as well as its accessibility from longer distances trough the near-by highway make the v-mall a 24 hours a day experience.

To enforce the open character of the v-mall, all enclosed spaces are designed as a metal-glass structure. During the day the v-mall invites its visitors with openness and public paths, at night the complex achieves its identity through its glowing appearance.

A computer-animated machine designed to measure the intensity of various impulses was given data about the flow of population and traffic during the 24 hours of a day in downtown Los Angeles. Freezing this machine at a frame where the demands for urban connection were best met created the distribution of the commercial and public spaces, as well as the in-between zones. From there the results of the machine had to be examined and interpreted trough more

美国购物中心在过去 50 年中的发展呈现出两种形式："超级商业区"与"中央大街"，这两种形式分别盛行于东西岸。两者间的差异主要是由于文化与气候的影响："中央大街"式的购物中心仿照美国市郊的小尺度—这样的场景淡化了纯粹的商业机能，让访问者暂时忘记现实；安置于停车场缓冲区域当中，超级商业购物中心有着相似意图——一个后现代的商业要塞。两者虽都依附于人造的公共空间，却都漠视其身处的公共环境。

V-mall意图融合商业与公共空间，成为可以强烈反映与都市环境关系的复合体。它周围的每条街均步行可及。一条连接内部与外部的都市小径蜿蜒其中。电影院和停车场位于低楼层，而主要公共空间位于自然光充足的上部楼层。公共的开放空间成为由社会行为的自组力（而非其建筑）定义的"都市压缩机"。所以消

相同级别的情况下进行。零售商与24小时开放的 V-mall 两者互惠。

封闭的空间用玻璃与金属的构造强化其开放性。最终发展成模糊了销售、娱乐与公共空间三者间界限的都市地景。访问者置身其中将可同时察觉都市环境与 21 世纪的资本主义。

现代的电脑辅助工具以启发性工具之势启蒙了建筑设计过程，本设计方案自最初到最终成图都运用了电脑辅助。 V-mall 是个试图探讨建筑数码时代潜能的方案。[翻译 蔡咏岚]

AFFILIATION: Nonlinear Architecture, New York
COUNTRY: U.S.A. / 美国

traditional ways of architectural discourse: envelope, space, circulation, structure etc. the v-mall in its final stage was developed into an urban-commercial landscape blurring the borders between retail, entertainment and public space. The visitors stay aware of the urban context and the social reality of 21st century capitalism without being fooled by the means of the main street typology. And like in a snowboard 'air' experience, they arrive at points of security and spaces of pure disaster.

The application of modern computer technology and new software with their abilities of 3d modeling, animation features and rendering capabilities are opening up a new understanding of the architectural design process. The v-mall project has been accompanied by the use of computer technology from the very beginning, by developing an abstract conceptual tool, to the final presentation in renderings and 2d drawings. Architecture gets pushed into a new era by keeping the computer involved as an inspiring tool throughout the entire design process, and even during construction. Architecture gains self-organizing potential, in which the architect does not lose control over the design but acts as a conductor. The v-mall project is an attempt to research the potentials of the digital age for architecture.

↓ Abstract Machine / 抽象机械

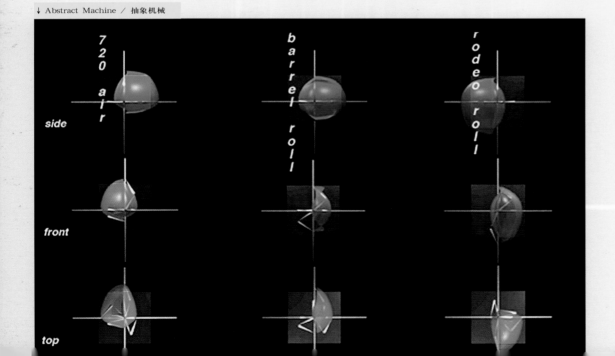

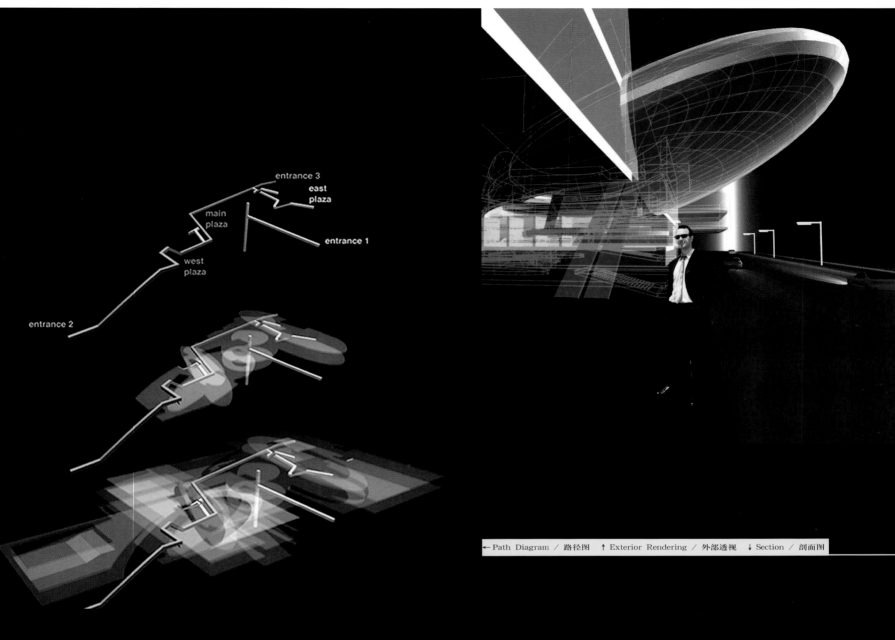

entrance 3

east plaza

main plaza

entrance 1

west plaza

entrance 2

← Path Diagram ／ 路径图　↑ Exterior Rendering ／ 外部透视　↓ Section ／ 剖面图

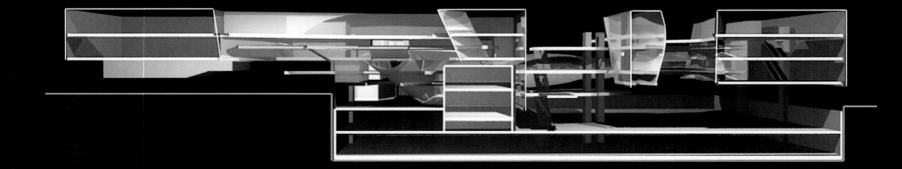

Analog Digital Process [adp] / 类比数码过程 [adp]
DESIGNER: Shawn Douphner + Fred Holt + Paxton Sheldahl

[Compression]

Analysis, selection and interpretation. The physical products of the analog become digitally compressed; a flattening of spatiality. Analysis, selection and interpretation follow. The digital then becomes the physical once more, a perpetual process. A process which allows the designer to re-interpret the everyday. A process which allows chance to manifest into space. With this process the interior of a refrigerator is manifested into spatial poetics.

[Object-field re-interpreted]

Our understanding of physical relationships between objects that our vision encounters is developed through experiences. That is we learn to interpret the everyday physical world through these experiences of objects within their fields. Yet, what if we could set new precedents for the way we interpret the everyday physical world. Rather, reinterpret the physical world of the everyday, such that we create new spatial constructs through the re-interpretation of objects within their fields. The analog, digital process allows this re-interpretation of the everyday physical world. The adp compresses, flattens our understanding of object and field into one; object and field superimposed, blurred such that a re-interpretation of our everyday and it's physical world allow the exploration of spatial constructs.

[Project / Process]

A project presents contextual (that which is) and conceptual (that which could be) information. That is, site to programmatic information. This begins to form a selective idea of what the project conceptually could be. The designer begins to reinterpret the contextual, conceptual information. The process of design finds its genesis.

The analog, digital process [adp] uses this technique; contextual, conceptual information and its re-interpretation. Its purpose is to reinterpret that which we believe we understand in order to explore various spatial constructs. The goal of the adp is to reinterpret the object field relationship as we understand it in the everyday. This flattening and then its re-interpretation of the object -field allows the exploration of spatial constructs. Thus, spatial constructs which are not derived from an autonomous idea or conception, but rather dependent on the contextual, conceptual information which act as the genesis of the project.

In order to show how the adp allows the re-interpretation of that which we assume we physically and spatially understand, an object that one interacts with everyday was chosen. That is, the means for exploring the creation of spatial constructs through the analog, digital process [adp] was that of the interior of a refrigerator. Using the adp, the goal was to take this everyday object and experiment with its potential to create/render spatial constructs. Using the adp, spatial constructs are explored through seven steps.

Step 01   Analog Space Capture: photographs of the spatial nature of Refrigerators taken. These analog space captures evoke the idea of space within a space. In this step, design is the transcription of a spatial experience by selective framing and interpretation, in order to flatten object into field.

Step 02   Digital Space Capture: The analog space capture photographs are transferred to transparent acetate sheets. The acetates create a spatial quality that allows the designer to see overlapping images creating a conceptual space for media experimentation and playful discoveries. Thus, the acetate sheets are superimposed upon each other to create experimental combinations.

Step 03   Analog-Digital Schemas: A series of critical and analytical diagrams from a selected digital space capture are created. This is the beginning of re-interpreting the object, field relationship. The seeing and making of schematic diagrams facilitates the understanding and interpretation of the spatial construct embedded in the digital space capture. In this exercise, the process of diagramming allows the re-interpreted object, field to create/explore new hierarchical relationships, and thus potentials for new spatial constructs.

AFFILIATION: Cal Poly San Luis Obispo
COUNTRY: U.S.A. / 美国

压缩分析、选择及诠释

类似物质进行数字化压缩，空间也被拉平，接下来是分析、选择与诠释。数码再一次变得具体，成为一种永久性的过程。它允许设计师重新诠释每一天的生活，允许偶然性在空间里的得到体现，既使是冰箱内都富于诗意。

物体——场所 重新诠释

我们对物体间关系的了解，是通过经验来发展的。那是我们通过体验物体以及它们的场域来学着诠释每一天的物理世界，如果这样的规则预先受到修改，物体与场域叠合了呢？

计划／过程

计划代表涵构与观念，像是空间计划。因此可在观念上了解计划。空间结构体不再是从单单某个构想或观念里冒出来，而是依生于设计方案原生的涵构与观念信息。以冰箱为例，使用adp工具，目标就在于通过对这个日用品的使用以及对其的挖掘去创造及体现空间结构体。让我们以下面的步骤进行探索：

步骤1　空间性质的照片
步骤2　将照片转成描图纸
步骤3　一系列评论与分析的简图
步骤4　以3D软件来建构模型
步骤5　3D模型是由2D图像建构而来
步骤6　以模型软件版为研究空间的工具
步骤7　最后决定使用的模型[翻译 赵梦琳]

Step 04　Digital Templates: After using the designer's tool of selection through analog digital tracings allow the experimentation of digital constructs begins via 3-d modeling software. The acetate combination and schematic diagrams are imported as underlays into formZ in order to diagram in different layers and line colors to represent different systems. Thus, the images are re-interpreted once more. In this step then, design is the selective re-interpretation of the composite image.

Step 05　Digital Reliefs: Based on the digital templates, three-dimensional objects are created from the two-dimensional explorations. A digital projection of the templates transforms this step into a three-dimensional study. The extrusion of linear forms provides a transitional study between the two-dimensional and three-dimensional worlds. In this exercise, design is the creation of three-dimensional objects derived from analytical, diagrammatic templates.

Step 06　Digital Spatial Manipulation Device: The modeling software is now used as a spatial manipulation device. Variable relationships between the digital reliefs are explored in virtual space. A selected number of composite model configurations are created as an interpretive transformation of the relief studies. Thus, the new object field relationship is derived from the original contextual, conceptual information. Therefore, allowing the re-interpretation of the everyday object, the refrigerator.

Step 07　Digital Spatial Insertion: Furthering the idea of selection, a final composite model was inserted into a defined context. In this case, selection was used as a way of compositionally arranging this new model in the given field, downtown Los Angeles. The digital model was spatially scaled to match that of the downtown. This insertion allows the re-interpretation of the everyday to be projected as a new possibility in an everyday space.

↓ City context view (Digital spatial insertion) / 城市涵构

↑ Street level view (Digital spatial insertion) / 街道层视景   ↓ Left side view (Digital spatial manipulation) / 模型左方

Digital relief / 数码模型 ↓

Hyper-sequenced Territories Cinemap / 超序动态领域

DESIGNER: Kenneth Ho

cinemap　The term CINEMAP was coined by Pierre Levy in his book, The Collective Intelligence: Mankind's Emerging World in Cyberspace. It is an organizing principle for organizing. It is the concept of a navigational instrument for organizing thought, amidst the era of digital media, recognizing the interactive implications of cyberspace, or this current informational universe in Levy's terms.

This term is the inspiration for the title of the thesis's investigations, CINE-MAPPINGS. CINE-MAPPINGS outlines a methodology for describing event territories through hyper-sequences. "Cine" in this case refers to the cinematic or the mode of representation of events in a moving sequence of images, and establishes the thesis's interest in SEQUENCING.

'Mappings' describes the act of marking or identifying territories, staking out the range of an event. It establishes the thesis's interest in the idea of TERRITORY.

territory　The notion of territory encompasses both the ideas of ground as well as the activity or event occurring on it.It is not just formally a surface, but a surface charged with possible and existent events, a multi-layered topography.

sequence　This thesis is interested in both the idea of sequencing as a mode of representation as well as a manner of describing an event. This means that the event is not construed as a one-off entity but is multi-valent, and is thus more accurately described as an EVENT STRUCTURE, consisting of a/many sequence/s of events.

The nature of the sequence in respect to an EVENT STRUCTURE or flux event, is neccessarily non-linear, and contains keys which causes the sequence to take different routes, resulting in multiple scenarios, resulting in a HYPER-sequence.

PHASE ONE　in relation to digital media, this thesis was specifically interested in the implications of animation techniques on modes of representation in architectural diagramming. Therefore this phase set out to establish a notational system for describing a multiplicitous event.

This notational system was effectively a script to be translated into an "animate-able" digital model. A digital model which represented the territories described in the event as 'evolutionary' surfaces.The multiplicitous event chosen was the ten cities described in the first chapter of Italo Calvino's Invisible Cities.

The first step was to describe these cities as a script. This involved a set of three 18x22" drawings, each in relation to a specific aspect of an event. The 3 aspects are described below. Bernard Cache describes the three basic conceptual elements used to describe a visual structure: [1]the frame (the square);[2] the inflection (the curve); [3]the vector (the arrow).

In the interactive event score, three similar concepts are used to graphically represent the multiple relational structures: [1]the durational frame; [2]the territorial score; [3] the vector score.

PHASE TWO　In phase two of the thesis, the damming of the 3 gorges dam was chosen as the site for a project, employing the techniques established in phase one.

In the event structure of the 3 gorges dam project, 3 durational structures have been identified as existing conditions.These 3 durational structures are the intervals of time that can be associated with different motions, different processes, and different trajectories at different spaces, that was described by AYSE ERZAN. *(Anytime. p102). They are also in this context, described as TEMPO MAPS.

"CINEMAP" 始见于 Pierre Levy 所写的《集体智慧》一书中，此名词是组织的思想原则，尤其是在数码年代。

"CINE-MAPPING" 即为利用超程序来描述事件的"界"的一种方法。"MAPPING"主要用于标记或定义"界"。

界域　不只是指某一范围，也指该范围的活动，它的地形学含义是多层次的，不光指地表，也包括地表上的活动。

程序　程序为描述事件顺序的方法，事件并非单独存在，而是有多种意义和价值的，不如以"事件结构"来描述它，它包含许多程序。

阶段 1
阶段主要是以电脑技术来模拟建筑物的空间流程，建立一种描述事件的系统，包括：[1]时间框架；[2]界的范畴；[3]向量范畴。

阶段 2
在此阶段我们利用阶段 1 的技巧选择三峡水坝做为基地。

阶段 3
"Mapping" 不同的地形学。

每个站均依两种主要时间尺度来建构：24 小时和 10 年。[翻译 赵梦琳]

AFFILIATION: Southern California Institute of Architecture
COUNTRY: U.S.A. / 美国

The 3 durational structures characterize 3 specific types of durations, or qualities of time unique to the Three Gorges dam event: [1] the archaeologic; [2]the transitory & [3] the pilgrimage. These 3 categories are also referenced by three found objects at the site: an imperial seal (the archaeologic) / the junk vessel (the transitory) / a temple barnacle, Shibaozhai. (the pilgrimage). These latent histories attached to the lost artefacts (cities, temples, archaeologic remains), are evoked with 5 stations, which real events and places. The station not only manifest symbolism, but they also acquire new meanings when a traveller arrives, a function emerges. station for PASSING / station for DREAMING / station for INSCRIPTIONS / station for RISING / station for RE-EADING,-COVERY

PHASE THREE

[1]Mapping differentiated topographies: Each station is structured according to 2 main durational scales: a 24 hr daily scale, and the ten year duration concurrent with the third final phase of construction of the 3 gorges dam. The end of the 3rd phase of construction heralds the normal operation of the dam, and the complete inundation of the cities along the part of the Yangtze that will become the reservoir. These 2 timelines represent the dual durational status that each station has to sustain. It's spaces respond to events that will occur within these 2 durational scales. The sequence of their organization is determined by the anticipation of possible events taking place within the 2 scales.

This dual status describes the differentiated nature of the stations. Each station engenders a differentiation spawned by a daily durational scale, the ten year evolution scale and, references to historical events. Each station is itself a differentiated topography addressing the past as well as the present/future. Together they also form a larger differentiated topography which anticipates evolution.

[2]Graphical grammar: Event spines—The event spines of each station describe the trajectories of each of their implied programs. It is a mapped lined generated by a traveler.

[3]Durational projections: All 5 stations are based the 2 main durational scales described above: a 24hr daily scale / a 10 yr evolution.

[4]Territorial profile: The profiles of each station describes the changing condition of ground in its location, reflecting the existing terrain and inserted programs.

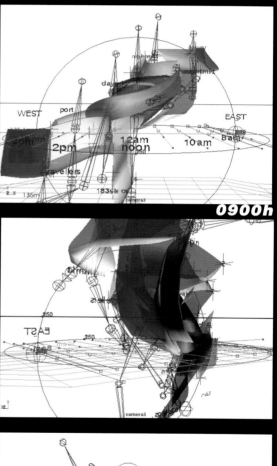

↓ 3d modeling of territories described by Marco polo, the traveler and respective natives in 5 cities – each over a 24hr period. (Phase ONE: Mapping Of Territories Of 10 Cities In Calvino's Invisible Cities – Chpt One.) / 根据马可·波罗的描述构建的3D模型，五个城市中，旅行者与当地的居民时间为24小时（阶段一：Calvino's "看不见的城市"中的10个城市）

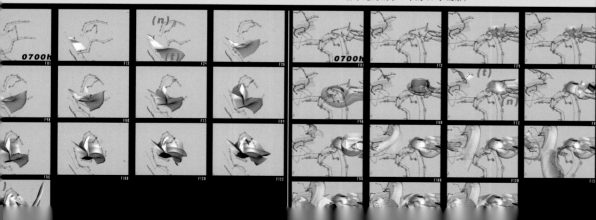

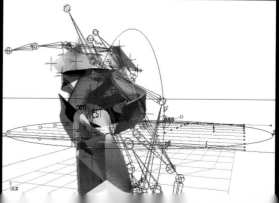

↓ Architectural representation as an interactive interface – form resu
between the traveler and the native. (Phase ONE: Mapping of Territ
Calvino's Invisible Cities – Chpt One) ／ 以建筑来诠释互动的界面(I

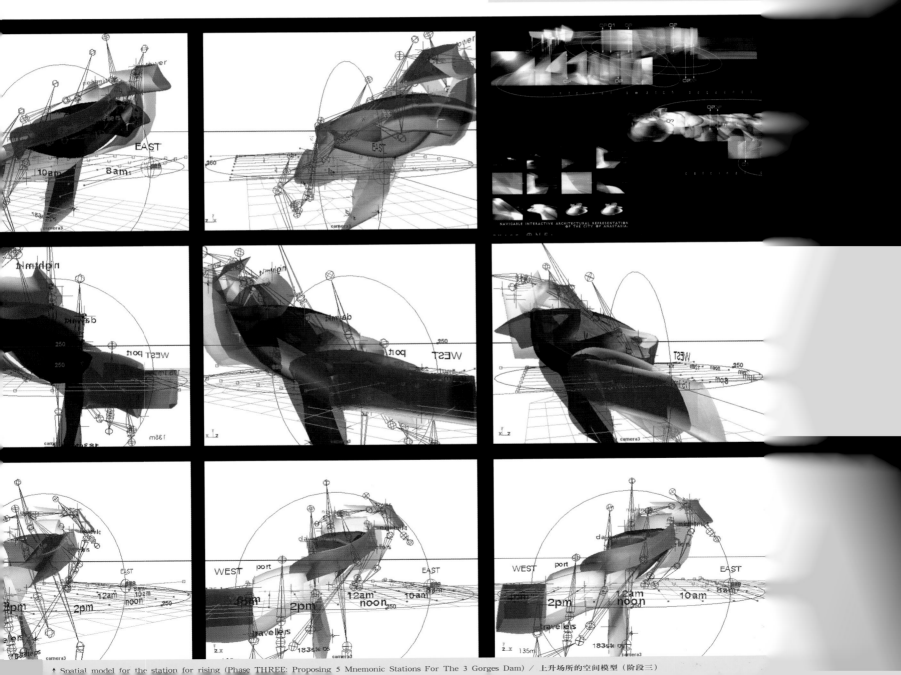

↑ Spatial model for the station for rising (Phase THREE: Proposing 5 Mnemonic Stations For The 3 Gorges Dam) ／ 上升场所的空间模型（阶段三）

Body is Environment — An Investigation into the Interaction between the Body and Its Environment

"一身" 一世界，一花一佛陀　身体与环境的互动

DESIGNER: Andreas Karaiskos

The work investigates the body analogy in architecture particularly through the investigation of boundaries. The nature of boundaries; whether they are permeable or impermeable is explored in order to prove the hypothesis that Body is Environment, that the two are so inextricably interwoven, that there are no separating or defining boundaries or edges. Boundaries are blurred and confused.

Perhaps the average person does not imagine himself or herself to possess a border or body boundary. The skin marks the edge of the body but many forces can help to enhance or disrupt our body boundary assurance and consequently affect our awareness of personal space. The notion of the body border can be associated with our physical environment also. We interact with various forms but perhaps never consider how our environments influence the body's movements and its physical and psychological being.

Boundaries in our physical environment are set up and influenced by society, as it must establish control over the individual. They are therefore, political and cultural divisions that attempt at controlling the movements and psyche of the masses. The visual penetration and authority of a surgeon/doctor/gynaecologist all involve a resignation to authority and power on the part of the individual who has little or no choice. It is intended that through the enhancement and renovation of the body social convention will be opposed. Through the digital penetration of the body and space, architecture evolves. By taking a particular site - Greyfriars doctor's surgery in Hereford established conventions are broken resulting in the freeing of the individual.

Greyfriars Surgery - Hereford, UK

The Greyfriars Surgery has been constructed on the remains of the medieval city wall adjacent to one of seventeen bastions, which is also the last remaining. Stones from the wall were removed during the initial construction of the surgery and were used as building material and incorporated in the construction of the basement level. The surgery has grown over the wall and now penetrates it. The following describes the principal issues:

[1]Technology: the tool of surgery - method of investigation and diagnosis
The individual becomes connected to technology in many ways. Technology enhances our senses and physical capabilities. It becomes the means by which the self is decentred. Multiplicity, fusion between the organic and the inorganic, between man and machine, between organism and environment, between self and other, is made possible through the introduction of technology. It is effective on both a macro- and microcosmic scale. The boundary between the 'real' and the 'non-real' becomes blurred.

[2]Movement: through public & private spaces, waiting areas to doctor's office
Today's urban form is based on moving bodies as our contemporary city has been designed to enable the free movement of its individuals. The increase of movement has been possible due to technological advances in transport systems. They enable an increase in our ability to travel both quickly and unobstructed. Modes of travel place differing demands and requirements on the environment while adversely 'liberating' architecture from the limits of urban context and human scale. The speed of the car and pedestrians vary, as needs overlap and contradict.

[3]Vision: surgeon makes his diagnosis based on visual penetration and scanning
Vision is the means by which the perceiver and the perceived, the subject and the object are synthesised. Internal space is united with outside space. The surgeon's gaze an intrusion into the patient's internal spaces. The division between the two is not distinct. The spatial divisions between the public and private realms have gradually become confused with the use of glass that is analogous to a permeable boundary. Solidity is the condition of impermeable barriers and boundaries.

AFFILIATION: Architect
COUNTRY: U. K. / 英国

身体与环境的类似性，尤其是界限—界限是否能打破，关系到 "身体即环境" 的假设。二者间的纠葛使这条界限模糊不清。

[1] 技术：手术的工具——检查与诊断

个人与科技和技术越来越息息相关。科技强化个人感官与体能，自我因此而扩张。藉由科技各种重叠融合因此发生：有机无机的模糊、人与机器关系的密切、有机主义与环境的融合、自我与他人的交叉。尺度无论大小，现实与非现实的界线也因此模糊。

[2] 动作：公共空间与私密空间；诊所的等候空间

现代的城市让市民愈来愈能随心所欲地移动，城市的造型也是基于此原则。市民能畅行无阻也应归功于科技，交通模式改变环境，使建筑空间免于城市的组成与人的尺度的限制，限制取消，人车速度因此改变。

[3] 视景：医生依赖视觉上的检查、扫瞄来诊断

视觉是用来综合看与被看，主体与客体的，网络则用来连接外空间，就如同医生必须洞悉病人的身体，二者非常相似。玻璃就像是可穿透的界限，它的使用会模糊室内外的界限，实墙则是不可穿透的界限。

[4] 界限：影响我们的行动

界限无所不在，影响我们的行动。最初它们是不可穿透的，包括垂直和水平的，我们也须依其位置调整我们心中的意象，如何界定这些界限取决于我们自己的经验，所以我们的洞察力将影响我们的决策，设计程序中必须考虑到这些界定空间的外延边界。

设计重点是环境与身体间的关系，界限的本质，强调了环境与身体间关系的复杂性。界限的本质复杂且可穿透，其交接处蕴藏许多可

[4]Boundaries: influencing our movement

These define our movements and are everywhere. They initially appear as impermeable however; they can take any configuration both vertically and horizontally. We must therefore constantly reconfigure our mind's image of space and boundaries according to their location. Our experience is subject to our ability to define these limits. Therefore perception is critical to our decisions. These boundaries and edges that define the extent of spatial requirements are to be considered in this design programme.

The movement of patients and staff at Greyfriars were analysed in order to establish the varying use of public and private spaces. This analysis represents the conventional use of Greyfriars where its size and layout has evolved from the demands of its occupants, following the conventional programme Form Follows Function.

To achieve a re-inscription of the body the movement paths of the users are used to define the limits of the various spaces. In this way these movement diagrams of the body become central in achieving the body's re-inscription into architectural form. The movement diagrams were modelled and extruded to give a three dimensional representation of for example circulation on stairs, or a patient lying down. A three dimensional model of the building was produced and an exercise of morphing the movement paths with its corresponding floor model was carried out.

The research has been concerned primarily with the relationship between the body and its environment. The work has through the exploration of the nature of boundaries, attempted to illustrate that there is an inter-relatedness that is so complex that the two cannot be separated. The nature of boundaries has been critical and has led to the conclusion that they are permeable. There is a point of intersection and overlap and I have been seeking to discover what occurs at this point of intersection. The morphing images attempt to illustrate what can occur at that point of intersection.

↑ Body Transformations 02 – Explorations of the Body Border. ／ 人体转化 2
↓ Greyfriars Surgery, Hereford – Plans, Sections, Elevations and 3D Model. ／ Greyfriars 诊疗室平面图、剖面图、立面图及 3D 模型

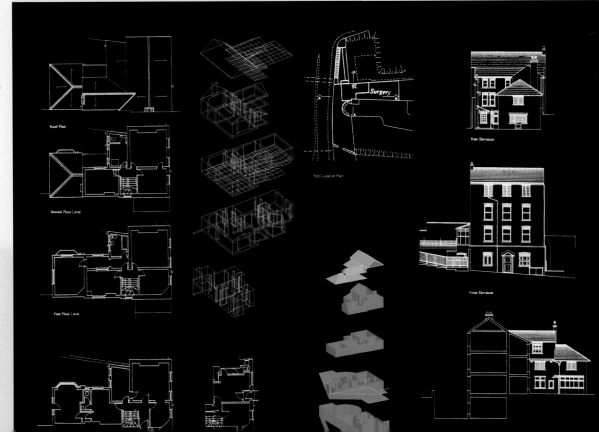

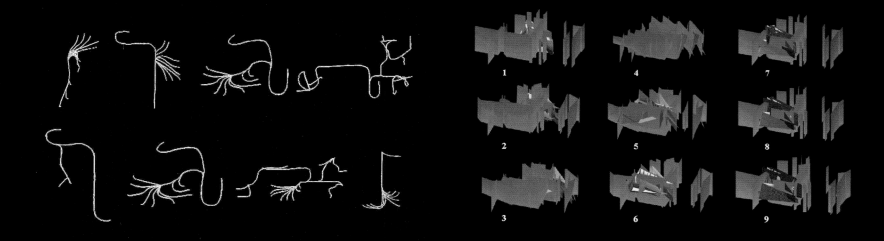

↑ Sketches Illustrating Movement Paths of Users and Occupiers. ／ 素描显示使用者的动作　Morphing Sequence 01 – Basement Level. ／ 渐变 01—地下层
↓ Morphing Sequence 02 – Ground Floor Level. ／ 渐变 2—地面层

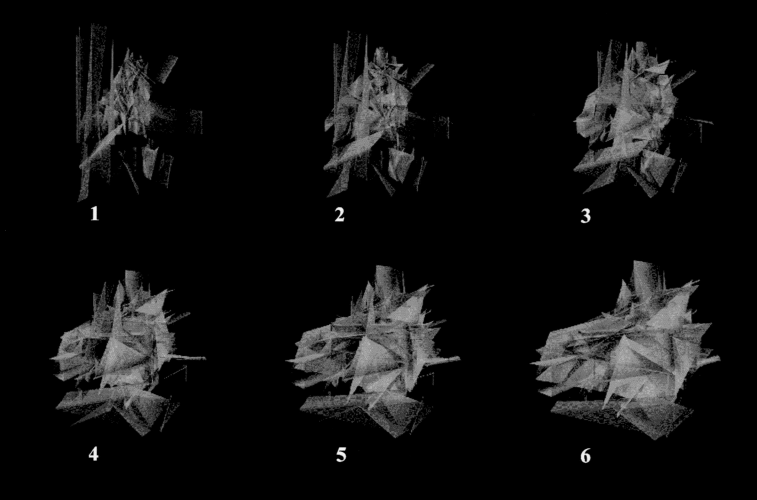

Animate Building / 动态建筑

DESIGNER: Chiafang Wu + Stephen Roe

At the same time that architecture is becoming "digital" through the use of computers in the design process and the digital is becoming architectural through the increasing use of 3-D simulated environments to understand & navigate digital information space, the digital and architecture are being invisibly integrated in a process that is not even apparent to most architects. The apotheosis of digital culture today -the World-Wide-Web- is serving as the practical model for the organization of all the processes which go into making a complex building today.

The phenomenon that is the Internet is an example of a complex self-organized system made up of many independent systems which was not planned in any meaningful way but arose spontaneously from a few very simple basic elements: [1]The IP (Internet Protocol) Address (eg. 209.174.311.6). [2]The DNS Server Address [3]The Routers which use these addresses to direct information.

The Internet is basically a multivalent directed flow of information, what makes it so powerful is the fact that every single piece of information (or IP packet) flowing over the web has a designated IP address which it must reach. A series of routers which points it in the right direction ensures that it reaches its destination. This means that every point on the net can contact every other point without fail (if not blocked by some superimposed security system). Around this basic principle a huge and virtually limitless network has emerged. But it is also around this system that the buildings of the future will be designed.

On a spinning rock of limited resources but rapidly increasing population, efforts to reduce the cost (both in terms of economics and resources) of building are receiving more and more attention. As our world develops from a manufacturing to an information economy the processing of information becomes cheaper as the processing of materials becomes more expensive (Norbert Wiener). Because of this we are beginning to see buildings which utilize ever more complex building management systems, controlling heating, lighting, ventilation etc. ("Smart Buildings") to reduce their energy consumption. Also with increasing emphasis on embodied energy smart structures will allow for lighter construction by responding intelligently to physical forces (this technology is already being used to counteract seismic forces on buildings in Earthquake zones). All of this creates large amounts of information which must be processed and distributed in a controlled way. A constant and directed flow of information is the result and the emerging standard for achieving this is the TCP/IP protocol which is responsible for the World Wide Web. In fact many buildings already integrate this technology, often without the architects even being aware.

Right now this technology is being used in buildings in a fairly conventional way to obtain information from sensors and use that information to control the various devices which make up the environmental control system of the building (effectors). But as buildings become more complex and structural and programmatic as well as environmental information is incorporated into a holistic building management system, more and more complex information networks will be created and new tools will need to be developed to understand, predict, control and creatively manage the building: Some of these new tools have already emerged in the development of the Internet: Logical Network Maps, Internet "Weather" Visualizations, Information "Landscapes," Internet Topology etc. which should lead to a very different way of visualizing the built environment:

电脑在建筑上的应用令建筑数码化，3D技术在模拟与体验空间上的应用使数码技术建筑化。在不知不觉中，许多建筑师的作品建筑与数码已合二为一。万维网—今日数码文化的典范，正是管理所有操做程序的实用模式，由此我们构建了更为复杂的建筑。

网络是一种复杂的自我组织的系统，包含许多独立甚至无意义的系统。它们其实是源自于非常简单的元素：[1]IP；[2]DNS服务器；[3]用这些地址引导信息的路由器。

本质上网络集合了多方向的信息，它之所以强大，是因为每个网络上的信息都有它的IP地址，路由器会正确的指出它的方向，确定信息会到达的目的地，如此可确保每个信息正确地与其他信息相连(不会为安全系统所阻挠)，然后便出现庞大、无限的网站，未来建筑势必也会依照这个原则建造。

为了解本数码建筑，我们先想像一个"整合的界面"，在一个小型示例中，调整结构、环境与科技。建筑的内外空间的相互作用形成建筑的界面，正如我们的皮肤其实是身体与环境间的界面，它包含了许多层次，特别是外层、内层及中介层，这有着环境、空间计划及结构的差异。构造物内能换气的空气缓冲区也是必要的，正如体液能在皮肤上散发气味、降低温度。将体液换成空气亦可应用于建筑上，即应用空气调节建筑物的通风及温控。

人工智能系统可随时感知建筑物的变化，建筑物已脱离以前机械控制的模式，而进入"Wiener"所设计的网络模式。[翻译 赵梦琳]

AFFILIATION: Architect, roe+wu.new.york
COUNTRY: Taiwan,China+Ireland/ 中国台湾＋爱尔兰

We see a convergence of computer network "architecture" and real world architecture which is not just analogous or an inclusion in the design process alone, but is, in fact, the next technological leap in construction.

As a starting point for realizing this digital architecture we envision an "integrated surface," incorporating structural, environmental and information technologies in a smart construction which responds intelligently to influences both inside and outside the building acting as an interface between the two, in much the same way that our own skin mediates between our bodies and the environment. Like our skin the surface is made up of several layers, specifically an inside layer, an outside layer and an intermediate layer -where differences in environmental, programmatic & structural load are actively dissipated. A buffer-zone filled with air which "breathes," expanding and contracting as necessary.

Fluids are the essential distributors of information, energy and temperature regulation across our skin. We propose to use the building's most essential fluid -air- to achieve the same effect: providing structural rigidity along with ventilation and temperature control.

With "intelligent" systems which learn and adapt to prevailing conditions the digital, in architecture is moving away from the robotic, programmed mechanism model to the self-organizing, dispersed, cybernetic model predicted by Norbert Wiener (one of the originators of this technology) in the 1950's.

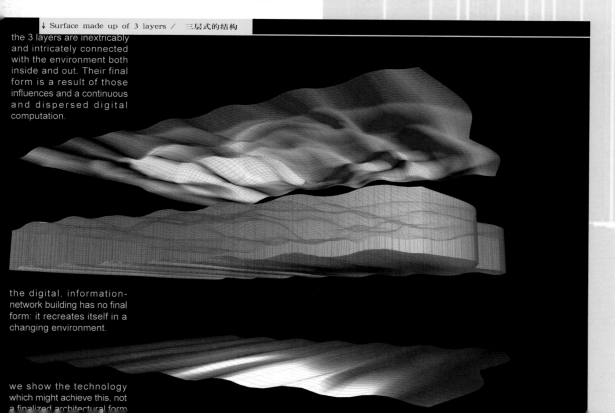

↓ Surface made up of 3 layers ／ 三层式的结构

the 3 layers are inextricably and intricately connected with the environment both inside and out. Their final form is a result of those influences and a continuous and dispersed digital computation.

the digital, information-network building has no final form: it recreates itself in a changing environment.

we show the technology which might achieve this, not a finalized architectural form

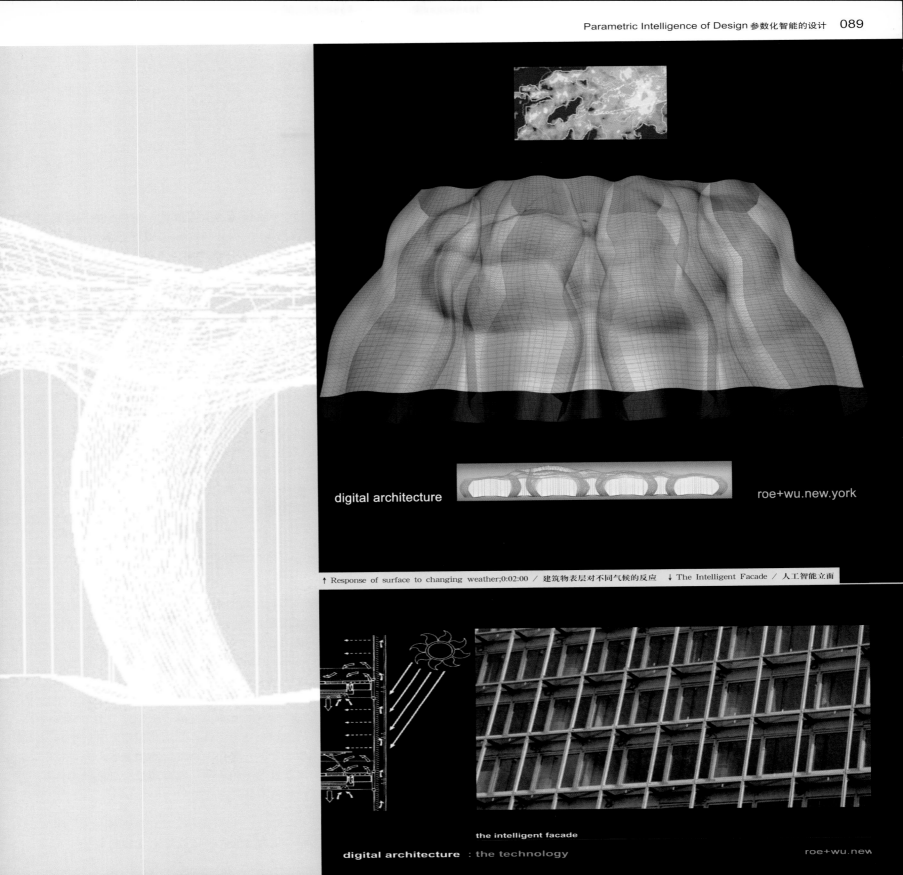

digital architecture

roe+wu.new.york

↑ Response of surface to changing weather;0:02:00 / 建筑物表层对不同气候的反应   ↓ The Intelligent Facade / 人工智能立面

the intelligent facade

digital architecture : the technology

roe+wu.new

Borgesian Borges Center / Borgesian Borges 中心

DESIGNER: *i.b.* **Bong**

Evolution proceeds through mutations, and...thought proceeds by fits and starts. Jorge Luis Borges has left so many fictions marvelous, imaginative, dazzling, inexplicable, ...and nevertheless very short. It is significant that his stories are short, for endless time and spaces are folded in them. What Borges tell me is, on the one hand, the fact that complexity or infinity can be embodied within the smallest entity of any kind and, on the other, we may not need to discriminate between real/actualized and virtual/potential phases of a thing.

As for Borgesian Labyrinth, now let's take Minotaur into our consideration. Unlike Ariadne's ability to interpret or Daedalus' ability to build, he is the only one who had the ability to live in the Cretan maze. It means that he had to construct almost infinite number of experiences at every moment in his life out of its singular finite structure where, he may have said, "each part... occurs many times; any particular place is another place".

Borgesian Borges Center aims at this kind of construction. There are two material models; Labyrinths and the Metro Cube. From the Labyrinths the intuition to endless time and space stems and Metro Cube gives the conceptual organizing structure. In this stage, accurate observation of the relations of parts, its notations, and diagrammatic analyses are main tools to draw a framework with. Then I can get a set of criteria for the primary unit to construct the whole. I would call it "Genotype". Genotype is merely the singular small object but contains in it almost infinite number of experiences as a virtual/potential form, which is obviously simple, but not a reduced simple form.

Finally the genotype mutates in accordance with the concrete programs and environments. Hence the plural combinations of Mutants.

→ **Mutants** / 突变体

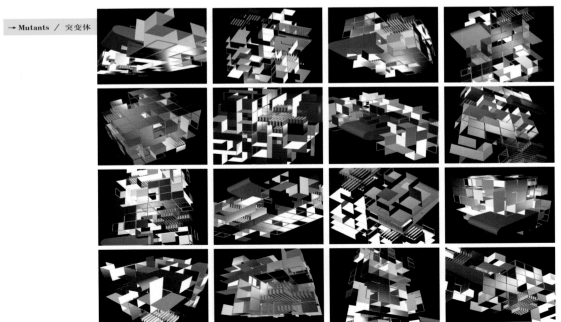

演化随着突变推进，而思想则间歇地进行。Jorge Luis Borges 留下许多令人惊叹、富于想像力、璀璨而又难以名状却着实简洁的小说。无尽的时间和空间交叠体现在他意味深长的短篇小说里。Borges让我知道，一方面，复杂与无限可以体现在任何微小的实体之中；另一方面，我们其实不用去区分物体真实的一面和虚拟的一面。

至于Borgesian迷宫，我们将Minotaur(希腊神话中的人身牛头怪物)纳入思考中，他不像Ariadne(希腊神)有诠释的能力或Daedalus(希腊神)有建造的能力，他是惟一有办法在Cretan迷宫里生活的。这意味着他必须从自身生命有限的结构中随时建产生于无限的经验，他自己也曾多次描述"每个部分……都发生过许多次，任何特定的场所可以是另一个地方"。

Borgesian Borges 中心以这样的建筑为目标。

它有两个物质模型：迷宫(Labyrinths)都会立方(Metro Cube)。在Labyrinths迷宫模型中延展了对于无限时间与空间的直觉，而Metro Cube提供了组织结构的构想。在这个阶段，对于组织结构部件关系的准确观察、标注记录和图表分析是建构骨架的主要工具。于是我得到判断基本单元的一组标准，由于可以构造出整个建筑，我称之为基因型。基因型仅仅是一个单一微小的物件，却蕴藏着作为虚拟形体几近无限的体验，它显然是一种简洁但并非简化的形体。

最后，基因型是随实体的计划内容与环境而产生突变的。所以它是突变体的负数结合。

[翻译 蔡咏岚]

Customizing Mass Housing: the Virtual Siza / 集合住宅

DESIGNER: Jose Pinto Duarte (Designer of the shape grammar) + Ana Runa Ferreira, Sara Eloy Rodrigues (Collaborators)
+ Alvaro Siza Vieira (Original designer of the Malagueira houses)

In the 1970s, Siza developed a system aimed at increasing user-participation in the design of mass housing at Malagueira. Devising implicit design rules, he used those rules to generate over 35 different layouts, ranging from one- to five-bedroom houses in an effort to incorporate the user's desire for a custom house into the design process. But limitations emerged and the system's potential to customize the dwellings was never fully used.

In an attempt to overcome those limitations, a shape grammar of Siza's design at Malagueira was developed. The grammar systematizes Siza's rules so that other designers can generate custom designs in the language. A new design by the author of the grammar°–the second author—was placed among existing designs and Siza—the first author°–was not able to distinguish it from his own designs. In addition, new designs by other designers°–the third authors—for specific clients also were considered to be in the language. A Web-based computer interpreter for the language is currently being developed. The interpreter is based on a mathematical model called discursive grammar. A discursive grammar includes a shape grammar, a description grammar, and a set of heuristics.

The shape grammar provides the rules of formal composition, whereas the description grammar describes the design from other relevant viewpoints. The set of heuristics is used to guide the generation of designs by comparing the description of the evolving design with the description of the desired house. Search is largely deterministic, which decreases the amount of time required to find a solution, thereby making it reasonable to develop Web-based implementations. When coupled with new prototyping and virtual reality techniques, the interpreter will enable users to visualize various design proposals.

20 世纪 70 年代，原设计者 Siza 在设计这个位于 Malagueira 的集合住宅时，曾尽最大可能的考虑使用者的参与程度，他发明了一套设计原则，发展出 35 种平面。依使用者的需求及其在过程中的参与，设计出从 1 居室到 5 居室的住宅，但因限于现实，这套方法从未实现过。

我们因此发展出另一套造型语言来克服这些困难，这套语言将原设计者的设计原则系统化，让其他设计者得以实现为使用者量身订做的住宅。第二代设计基于原有的设计，而且不易区分。而第三代设计是为使用者量身订做，并仍采用最早的设计语言。经电脑诠释的语言建构于数学模型上，我们称之为语言文法，包括造型文法、叙述性的文法与启发性的设计方法。造型文法目的在于提供造型的组合方法；叙述性的文法则旨在提供多方向的观点；启发性的设计方法则可比较不同时代的设计。寻找的方法具有决定性的影响，因此会影响找到解决方案的时间，而这也是运用网络平台的原因；虚拟技巧的运用，会让使用者设想不同的设计提案。[翻译 赵梦琳]

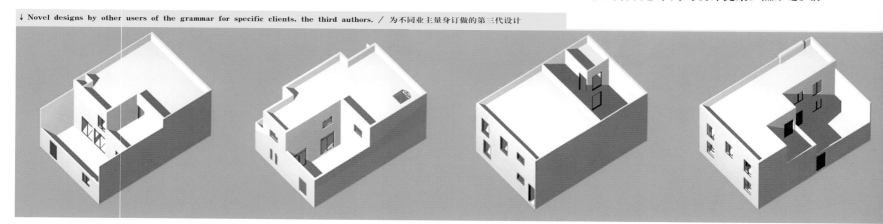

↓ Novel designs by other users of the grammar for specific clients, the third authors. / 为不同业主量身订做的第三代设计

AFFILIATION: Massachusetts Institute of Technology, U.S.A. + Laboratorio Nacional de Engenharia Civil,
Portugal (Ana Runa Ferreira, Sara Eloy Rodrigues) + Universidade do Porto, Portugal
COUNTRY: U.S.A. + Portugal / 美国 + 葡萄牙

Transitional Pod / 过渡空间

DESIGNER: Hunt McGarry Architects

Locked in our own time and space we"view the present and future through a rear-view mirror". (McLuhan paraph.)

While the rate of development accompanying digital technologies/cyberspace renders yesterdays hard/software obsolete, perception of our environment whether real or virtual appears to change at a far slower rate. McLuhan referred to still viewing the world within a Renaissance context, as a series of hierarchical objects in a perspective space.

The aim is to make a real space, using current technologies, that is analogous to the multi-layered, non-sequential, non-contextual environment of cyberspace and the dematerialization implicit with digital technologies. The Transitional Pod aims to make transitions between object and image, to dismantle the context/semiotics that distinguish between the real and the virtual. We are looking for a list of ingredients, clips, a specification, a mix. The Transitional Pod is a museum, a street, a house - it is all or any space - it could exist in a tepee or a dymaxion.

"被锁在我们的时空里，我们仅能通过后视镜窥探过去和未来" —— Mcluhan。

数码科技的发展，使许多电脑软硬件过时。人类对现实与虚拟的感知却相对慢得多。McLuhan仍以文艺复兴的透视性的、富于层次的方式来看空间。

本设计的目的主要在于用适宜的技巧来模拟一个空间，这个空间是多层次、非序列性、非涵构及抽象的。过渡群组功能在于转化物体与影像，拆解现实与虚拟之间的涵构与语意等，寻找成分、判断规格及组合。这个过渡空间可说是一座博物馆、一条街、一栋房子，它可以是任何空间，也可以在一顶帐篷中。[翻译 赵梦琳]

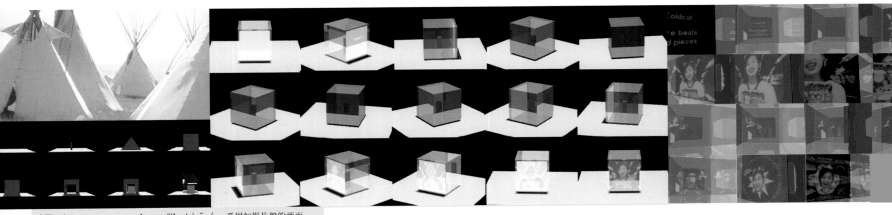

↑ The images are arranged as a "ilmstrip" / 一系列如影片般的画面

AFFILIATION: Hunt McGarry Architects
COUNTRY: U.K. / 英国

How to Increase Land Value with Your 1:1 Kit /
如何以 1:1 的模型增加土地的价值

DESIGNER: Stefania Kenley

The project addresses a particular site of Bucharest, but it can be relevant for neglected zones of any city. The insertion of a 1:1 kit in Bucharest's neighbourhood follows the logic of existing functional zones and their dynamic: accommodation, communication, food market, storage, etc. The web site leads you trough the feasibility of the kit as object and through the implementation process of its programme related to its specific site.

As object, the 1:1 kit would be an inhabitable, portable and flexible environment that you can build yourself to suit you and your life style. It involves an easy assemblage of a flexible structure made of detachable and interchangeable pieces. It has a rigid structure using a unique module with telescopic detail connection, which you can pack away; it is also wrapped in a textile piece with multiple zip connections that you can fold and ware as an overcoat when leaving.

The 1:1 kit can be built by anyone or distributed to everyone, but in this project it is relevant mainly when its miniaturised connection system has an impact on different functional zones of the site and on their potential extensions. The interchanging components allow a flexibility of use, so that an instant shelter can be inserted in the city network covering a variable number of people.

The kit can work both as object and programme only when it is accepted by the city. The acceptance of the kit as object would imply its connection to improved infrastructures which would be able to support the existing and which would allow new interventions. On the other hand, its assimilation as programme does not depend only on the quality of infrastructures, but also on recovered civic qualities, including an acute awareness of the feeble condition of the traveller, the foreigner, the marginal, or the homeless. Only when the inhabitant of such an alien kit can plug in the infrastructure and in the social system of a city, its urban space will become the shifting stage of a travelling world.

PS. This project has had the technical support of the University of Greenwich and of a.i. productions and it is dedicated to stefitokyo@hotmail.com for our e-mail exchange about the current condition of the "wanderer".

此方案位于罗马尼亚首都布加勒斯特，但它其实可适用于任何城市。这个 1:1 的模型仍遵循现有基地的分区及动感：酒店、交通、超市及仓储。参观者可通过网络体验模型来观察建筑计划与基地的关系。

模型其实是一个富有弹性的环境，可依你的生活方式调整。它由可拆解的构件组成，而模型结构则采用特殊可收藏的模块；它又可收在有拉链的袋子里，折起来就像是一件外套。

模型可给任何人用、可由任何人组成，但重要的是系统对基地不同分区的影响，模型可交换组合提供弹性，也可组成容纳市民的庇护所。

模型作为一个客体被城市所接受时才有现实意义。其存在可改善城市的公共设施及城市现况，又能提供新意。而此模型的空间计划是否有效，仍依靠空间及城市的性质及它能否考虑到一些弱势者：外国人、旅行者、边缘人、流动人口。如果此模型能被纳入公共设施，而居住其中的弱势者能为社会所接纳，那么它其实也会成为这个流动世界的变幻舞台。［翻译 赵梦琳］

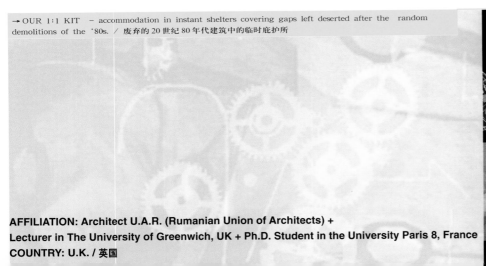

→OUR 1:1 KIT – accommodation in instant shelters covering gaps left deserted after the random demolitions of the '80s. / 废弃的 20 世纪 80 年代建筑中的临时庇护所

YOUR 1:1 KIT /

food market
6.00 - 18.00

sleep
21.00 - 9.00

swim
7.00 - 21.00

storage

AFFILIATION: Architect U.A.R. (Rumanian Union of Architects) +
Lecturer in The University of Greenwich, UK + Ph.D. Student in the University Paris 8, France
COUNTRY: U.K. / 英国

Digital Art Museum — City Life / 数码美术馆—城市生活

DESIGNER: Shen-kai Tang / 唐圣凯

Life design of the city From 1700 to 2000, the urban texture of this site actually grow. To predict the future texture in the coming 1000 years, we use the genetic algorithm form the computer science.

Transforming these urban textures from 1700 to 2000 into a couple bitmaps. We start to calculate these bitmaps with the genetic algorithm, which includes the reproduction, cross over mutation and evaluation. In the algorithm which include thousands of generation, we select two bitmaps as the base of the future texture.

From the texture, we extract the skin, body, structure and function as the architectural layers, as well as the design principles of the digital fine art museum.

Our design treat city as a biological existence. It grows, evolves, even decides its future consciously, providing a new possibility, for architectural design. [Translated by Moon-Lim Jau]

城市生命体设计 本设计是通过观察基地附近公元 1700 年到 2000 年的建筑物特点后，发现了一个类似生物生长的现象。因此便希望能够利用电脑科学中的基因演算法来作为设计的基础，进一步地来模拟未来1000年城市的生长特点。

首先，先是将观察到的公元 1700 年到 2000 年的城市特点转换成数张可供运算的点阵图档，再将这些点阵图档作为基因演算法的母体来进行运算。而其中运算的机制包括复制、交换、突变以及评估。在经过数千代的演化之后，我们取得了两张点阵图作为未来1000年城市建筑结构的特点。

我们再更进一步地将这些基地纹理进行建筑元素的分离与操作，从中我们解析出了皮、体、结构以及机能等数个建筑设计的层次，最后便完成了此数码美术馆的设计。

本设计旨在将建筑物或是都市视为一个生命体，它会生长演化甚至会有意识地来决定它们的未来，更进一步地提出建筑设计的新的可能性。

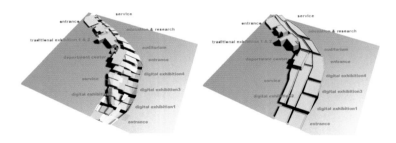

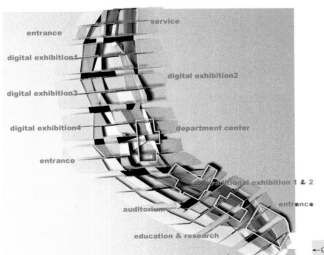

←Concept of architecture design / 建筑设计

AFFILIATION: Architecture Program, Graduate School of Applied Arts, National Chiao Tung University /中国台湾交通大学应用艺术研究所建筑组

NCTU—Sun Tracer / 逐日者

DESIGNER: Chao-Jen Wang / 王昭仁

Traditionally, there are wide varieties of imagination and needs between sunlight and architecture. The development of computer aided architectural design nowadays gradually involves into "design process" and generates different results The NCTU_Sun TRACER is a real project, a computer museum of National Chiao-Tung University and a new investigation between sunlight and architecture. The idea of NCTU_Sun TRACER is based on a workable scenario I set up, as follow "During summer solstice AM 10:00 to PM 2:00, sunlight doesn't stream into main interior spaces". And the project was exercised in one course of Virtual Design Studio. More precisely, with computation and simulation, sunlight can be controlled precisely when it comes to building. From initial concept to sunlight validation of interior spaces, I devided whole design process into four phases, which couldn't be achieved without a computer. A computer let us have precise power on sunlight control.

PHASE1: Site & Function　At the first stage, an sequare was choosen alternatively on a adequate position of the stepped site. Then following the contour and necessary functions of the computer museum, a rough digital model was built.

PHASE2: Sunlight & Form　Under the consideration of possible damage of display, a workable scenario was set up: "sunlight never drops into the main interior space between 10:00 am and 2:00 pm during summer solstice". The rough digital model where the position of sunlight projection was precisely recorded at specific points in time by a computer. Environmental data, such as altitude and latitude of the site, were also added in order to produce an accurate simulation.

PHASE3: Sunlight & Opening　After above steps, the proper sunlight requirements were determined, and these findings were applied when designing the openings of main structure.

PHASE4: Space Quality & Design Development　In the final stage, several computer images rendering by radiozity techonology were used to validate the lighting quality of interior spaces. These four design processes indicate that we never have such strong power on sunlight control which reflacting overall form without a computer.

I find most people couldn't realize the complex spacial relationship among such rendering images immediately. I tried to figure this problem out in a animation, a VRML model, or a traditional physical model. For some restriction of presentation, I choose the last solution and got good response from my audience. In addition, unfolding the complex volumn into 2D surfaces for constructing a physical model, seems just like precasting some panels in factory for real building procedure.

建筑与阳光之间向来存在着许多想像与需求，而今日的电脑辅助设计亦着重于设计过程。本方案是真实的案例——交通大学的电脑博物馆，设计源自夏至时光上午10:00到下午2:00射入室内的阳光，并应用虚拟实境技术。今天的电脑技术已可精确模拟射入室内的阳光，我因此将设计分为四个阶段，并利用电脑进行操作。

阶段1：基地与机能

先在坡状基地上选择一块广场，再以电脑建构基地地貌。

阶段2：阳光与造型

为避免展示品受到损害，需要营造这样的建筑气氛："夏至时阳光并不会直接照到展示品"，记录阳光照射位置，精确地建构电脑模型，包括基地的经纬度也需要计算在内。

阶段3：阳光与开口

经过这些步骤，可决定阳光的位置，并据此设计主体的开口。

阶段4：空间品质与设计发展

在设计的最后阶段，我们以3D电脑技术来模拟室内的光环境。综合四个阶段可看出电脑的重要性。

我发现这样复杂的空间关系并不能通过透视图而为人们轻易了解，因此考虑改用VRML或传统模型。在现实条件下我采用了后者并获得良好的反应，而之前所做的3D助于我构建传统模型。

← PHASE3 / 阶段3

AFFILIATION: National Chiao-Tung University, Institute of applied arts / 中国台湾交通大学应用艺术所

Morph-Forms — Digital Art Museum / 形态造形—数码美术馆

DESIGNER: Chien-tse Yang / 杨健则

Our city is full of the conflicts: the difference between the new and existing buildings; the global and the local culture. These conflicts are shown in the images of our surroundings, however, being transformed, these images then turned to be in harmony. The transformation is composed of various overlapping volumes, while the buildings around the site are full of these conflicts . All these are the elements in our design.

We use the "buffer", composed by the transformation, to moderate these conflicts. Here is the sequence: Assume the cultural center and concert hall as the basic elements for the transformation. Use the computer to trace the transformation as the principle of the moderation. Install the architectural program then mix with the transformation. [Translated by Moon-Lim Jau]

现今城市之中充满着冲突，无论是新旧建筑物之间的差异性，还是全球化与地区文化之间的势力消长，这些冲突直接反映在触目所及的影像之中，而拼贴之后却出现另一种和谐的转化。这种变化像是不同几何形体叠合后产生的综合体，而在基地周围的建筑物也存在着这种不协调的冲突，从而构成了得以进行操作的元素。

在本设计方案中，我们将"缓冲器"定义为用来缓和在不同形体之间的力量冲突，合并由形体变化的过程所组成。希望通过设置缓冲器的方式，使基地上所产生的冲突得以缓和，设置的操作顺序是：先将基地旁的文化中心与演艺厅视为形体转换的基本操作元素，使用电脑将形体变形的过程加以记录，成为操作形式的基础；然后将建筑的程序置入，使之与变形的形体相互融合。

↓ Approach of architecture program / 建筑空间计划

AFFILIATION: Architecture Program, Graduate School of Applied Arts, National Chiao Tung University / 中国台湾交通大学应用艺术研究所建筑组

Rulebased Housing / 规则之家

DESIGNER: Alois Gstoettner + Martin Emmerer

Natural structures and processes are determined by unshakeable rules. The human beeing is part of the game. To understand natural selforganisation one can not only cut nature into bits and pieces but has to explore its processes. Once you have studied the arranged bits and their mutual interaction the entire image of the puzzle occurs. human intellect can not solve the puzzle. even very simple interacting komponents can end up in a result that "blows our minds" (david bowie, major tom). The simulation of complex processes on the computer offers us the chance to get a clearer view on the image generated by (apparently) uncountable aspects. To transfer natural, dynamic systems of the "real" world into the abstract and artificial "cyber" world there is the need of an mathematical model's abstract language.

A special species of such a model is the cellular automata that is especially capable of describing networked activities of various subsystems. It has the remarkable quality of an ACTIVE MATRIX, by modern science seen as the key to understand creative, autonom, complex structures including biological life. The cellular automata describes the interaction of any number of elements which obey certain rules and determine each other in their local neighbourhood of their enviroment.

The elements of a cellular automata are a 3-dimensional grid of cells. Each cell has got a property. Adjacent cells are able to communicate and influence each others properties. The information one cell receives from its local neighbours determines its future properties. The exact influence of a property of a neighbour cell on the developement of a certain cell, their "way" of interaction, is defined by the RULES of the automat. The development of properties has a "beat": step by step the property of all cells change according to the given rules.

THE ATTEMPT TO WORK WITH COMPLEXITY PROVOQUES THE FOLLOWING: [1] AN ACCURATE EXPLORATION OF EXISTING CONDITIONS. [2]WELL DEFINED RULES FOR INTENSIVE INTERACTIONS ON A LOCAL SCALE. [3]A QUITE INDIFFERENT ATTITUDE TOWARDS THE OVERALL CONFIGURATION.
THE ACT OF DESIGNING SHIFTS FROM PRAGMATIC DETERMINISM TO DEFINING RULES AND PROCESSES, WHICH MAKES AN EVOLUTIONARY ARCHITECTURE BECOME POSSIBLE.

The cellular automata is a suitable model to generate new typologie of housing through its capability to inegrate complex nonlinear conditions and through its capacity to swiftly react to changes. Not mass and energy, but information is the cellular automata¥s raw material.

→ ALGORTIHM, A: …An algorithm incites the cellular automata by optimising the spatial "living units" according to the rules. This process is done through a programme that step by step transfers the existing into a more optimised state. The future property of each cell depends on the adjacent cellproperties and the local environmental parameters. Rules are defined as dynamic attractors. The system starts moving and tends to self organisation through a autonomous rule based process incited by mutual interaction of its elements and their environment. Starting with a homogenous field a not formal determined spatial structure emerges similar to an organism …/
试算法：试算法可利用细胞式自动化来找出最适合的居住单元，以一定的程序将现况转化为最佳的情形。未来的系数有赖于单元的系数及环境，而自动化的规则却为动态的诱因。系统开始移动，并自行组织，而自动化的规则则依赖环境与元素间的互动，由同质的场所开始，形成如组织般的结构。

自然界的结构有它的规则，人的存在也是其中的一部分。如要了解自然界的结构，我们必须先详细地了解及剖析自然界，了解了元素间的互动，整个结果便会如拼图般清楚呈现。人们的智慧不仅可透视这些拼图，更会激发心灵火花，电脑让我们从更多角度来设计更复杂的影像，将真实世界转化为虚拟世界，而不仅仅是抽象的数学模型。

细胞式自动化适用于描述不同系统的网络，它是积极的模型，也是可创造性的、自动化的、包括生物生命的复杂结构的关键。这个自动系统也善长于描述系统间的互动，这些互动其实遵循某些规则，并彼此影响。细胞式自动化是三维的格子系统，每个细胞各有系数。彼此邻近的细胞会互相影响它系数，从而决定未来的系数，至于如何影响则由自动化的规则来决定。依其规则，以其节奏发展。

程序的复杂性有下列特性：[1] 探索现况；[2] 从社区的角度来制定互动的规则；[3] 对整体的不同看法。

建筑的改革，即由程序化的方法演进成制定规则。细胞式自动化的非线性特点及对变化的快速适应力，很适合用来设计集合住宅，而信息就是自动化的基础，而不是量体或能量。[翻译 赵梦琳]

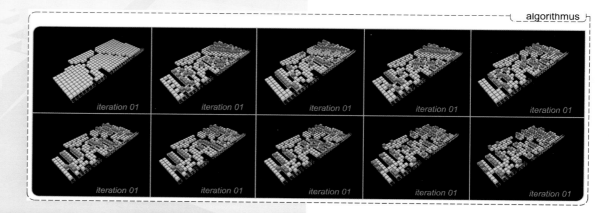

algorithmus

iteration 01 | iteration 01 | iteration 01 | iteration 01 | iteration 01
iteration 01 | iteration 01 | iteration 01 | iteration 01 | iteration 01

COUNTRY: Austria / 澳大利亚

Exploration of Digital Typology

Human civilization is a history of classifcation. Humans have ciassified natural objects to meet their every needs, thus putting objects into realation with each other as parts of this classification system. Different parts form into different combinations, and create a method of reading - an architectural typology. Architecture from Ancient Greece on has used a strict identification and classification system, and a similar system was developed in Chinese architecture in the East. Digital architecture, however, has created massive changes to spatial concepts. Space now may be virtual, not real; walls are no longer walls and floors may not be floors. The concept of a "part" is not as clear as it once was, while relationships between parts are increasingly hybrid. Does architecture no longer require types? Or has digital architecture, like other concepts before it, challenged existing concepts? Because this not only touches on design units, but also on building methods, this is destined to be an important direction of the future development of digital architecture.

数码类型的建构　　　　人类的文明就是个对万事万物进行分类的历史。为了各种需要与方便,人们会将自然物与人造物进行

分类,各分类系统中会形成许多不同的单元与单元之间的关系。不同单元形成了不同结合方式,并形成了长久以来对建筑建构

的一种阅读方式和思考方式。自古希腊以来,建筑就建立了一套十分严谨的分类与命名方式,以便建构建筑类型。在东方,中

国的建筑也有类似的严谨的系统。但是,接下来的问题是,数码时代的建筑在空间概念以及思考媒材与营造技术等方向都已产

生巨大的变化,空间也可能定义不明而且虚拟。墙不再只是墙,而地板也不再只是地板,所有"单元"的定义与界限不再清楚,

"关系"更是暧昧而又复杂。因此,到底是数码建筑不再需要类型了呢? 或者,数码建筑类型也会像其他概念一样,挑战长久以

来建筑所建构的类型概念,而自行建构一套建筑的数码类型。正由于这不仅涉及设计单元,也与营建过程与方法有密切的关系,

所以这是数码建筑未来发展的重要方向之一。

DynaForm — Cablecar Station / 动态形体—缆车站

DESIGNER: Kuo-chien Shen / 沈国健

DynaForm　DynaForm is a response between environments and events of the site, human activities and spatial sensibility. Instead of being a simpling or extension of cuts from dynamic traces.DynaForm is a mobile form based on human activities, expressing a context implicit in the site through the linking of events. It transforms these relations into force, field, and traces, as the references for design.

The forms of space were once commonly defined by functions of human activities, and, on the other hand, revised by the intervention of activities. The DynaForm blurs the step-by-step relations between forms and functions with the evolution of time, and breaks the one-to-one or hierachical constrains, which were replaced by the script in the computer set. The script introduces non-liner functions in which the parameters in the equations are related to the intensity of activities between events in architecture. The state of the form is controlled by the adjustment of parameters of the system. The low value correspond to stable behaviors, and the behaviors become more complicated when the value becomes higher. To the ultimate, it produces the chaos.

Besides creating references in non-liner system through description of topological relations of space and introduction of linking intensity of events by the digital machanism, the DynaForm defines references from the traces of human flows in the site and space. The traces are not the liner relation of two points,but stated as a spatial vector that implies force and direction. The vector leads the flows, intersection and deflection of different forms. The bodies and the forms collide with and adopt to each other. It can not be predicted what forms would be generated in the design process, but the shapes evolve with the design progress and interprets it simultaneously. The DynaForm attempts to make more intensive connection in the architecture events and environment, skin and volumes, forms and spatial sensibility.

Cablecar Station　The cablecar station is located in Hsin-Peituo, Taipei, Taiwan, China as an important transit station to link Peituo and Yang-Ming Mountains. The site is in an urban park, which is situated on a complicated traffic joint, and faces a traditional-formed NRT station on the opposite side. For this design, such an environment provide possibilities to deal with the interlacing of old and new, interior and exterior, and also to deal with the issues of 3 different levels, relationships between the building and the urban context, the surface and the volume, and the inner spaces.

[1]Volume: For the volume, by defining the functions and the circulation relationships of environment as dynamic forces, new type of interaction of each space are produced. This makes the spaces more dramatic and more powerful connection with each other.

[2]Surface: The relations between the surface and the volume is just as clothes and bodies, this indicates two levels, one is the relation between surface and the urban environment, which shows how surface reflects the environment and generate a new correspondent one. The other is the surface and volume, which create inter-spaces by the relative forces in between, and reinforce the interaction between urban park and architecture, outside and inside.

动态形体　动态形体的操作介于对基地环境与事件、人为活动与空间感知的回应，并非自由形体的取样或动态轨迹片段的延伸，而是基于人为活动所产生的流动形体，通过事件的连接传达基地环境所隐含的脉络，并将其关系转化为力量、场域及轨迹，作为设计参数的依据。

过去，空间形式一方面来自于人为活动所约定俗成的机能定义，另一方面则因活动的介入而有所调整。动态形体跟随时间轴的演化模糊形式与功能所存在阶段式的关系，也打破传统空间关系中一对一或阶层逻辑性的制约，取而代之的是存在于电脑机制中的表达式。表达式引入系统中的非线性函数，而其中运动方程式的参数值关系到建筑所存在事件间的活动强度。系统参数调节控制着形体的状态。小的参数值对应于稳定的行为，参数值越大，系统的行为也亦趋复杂，甚至导致混乱的状态，参数的范围维持动态形体的平衡。

动态形体除了透过电脑机制的表述式来描述空间拓扑关系及引入事件连接的强度作为非线性系统中的参数外，另一方面则是来自于人流动于场地与空间中的痕迹。痕迹在此被叙述为空间向量，它隐含力量与方向性，并非两点间所形成的线性关系。向量引导不同形体间的流动、交错与偏离，身体与形体在同一时空中相互碰撞与修正。在设计过程中产生的形体无法预知，形状的演变过程跟随整个设计的进行

AFFILIATION: Architecture Group, Graduate School of Applied Art, National Chiao-Tung University /
中国台湾交通大学应用艺术研究所建筑组

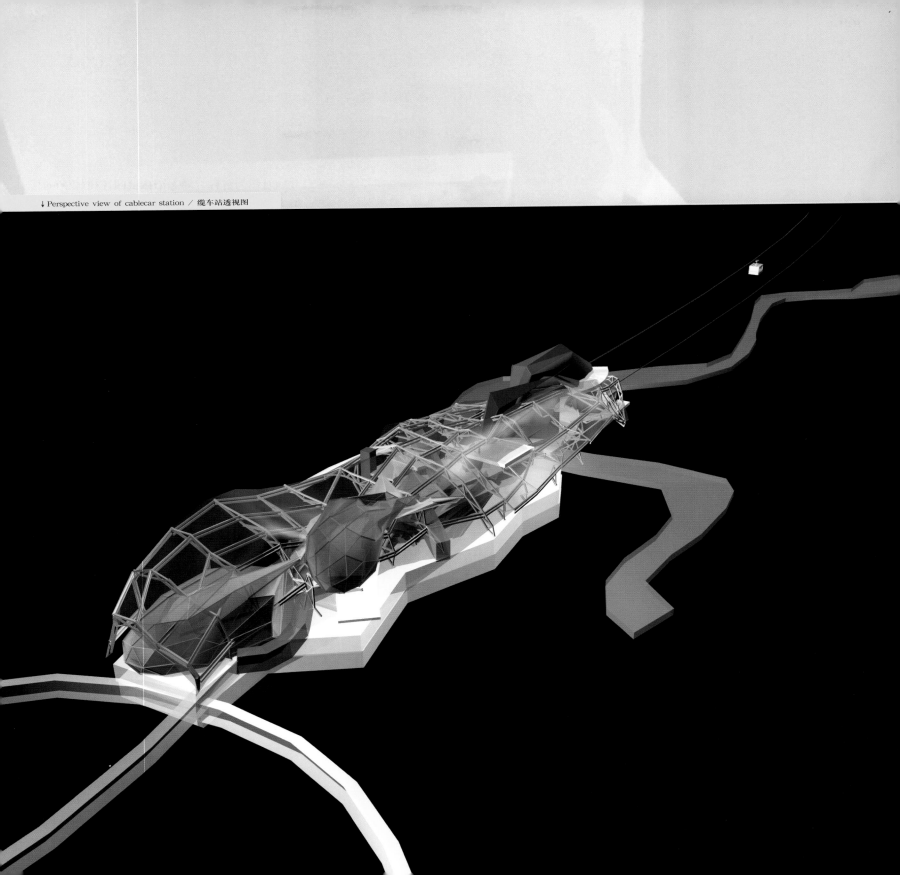

[3]Slabs: The creation of slabs comes from the lofting of circulation pace. The section lofts along the pace. The factors operated are defined by parameters, including width, angle, thickness, radian; and it becomes the way that spaces link each other.

[4]Structure: After the volume, surface, and slabs are created, 16 sections of equal span are made in the buildings. 16 structural frames are determined by the contour of sections with consideration to how the frames support the volume and slabs. After the 16 frames are created, each two sections would be linked by orthogonal frames to build the main structure.

并同时作为解释。动态形体试图让建筑空间中的事件与环境、表层与量体、形体与空间感知有更紧密的联系。

缆车站　　缆车站位于台北市新北投，它是连接北投与阳明山的一个重要的转运站。基地处于一个复杂的交通点上的都市公园里，侧面对的是一个有传统中国建筑造型的捷运站，这样环境为这个设计提供了一个处理新与旧，外部与内部复杂交接的可能性，凭借环境因素动态模拟来处理三个层次的问题，即建筑本体与都市涵构、建筑外表与建筑内部量体、建筑内部空间关系。

[1]内部量体：建筑内部量体经由机能定义与来自于环境的动态关系作为量体间互相作用的力量，去产生每个空间互动的新形态。这样的做法使得空间产生更强的联系并形成戏剧性的空间。

[2]外表：建筑外部的外表与量体如同衣服与身体的关系，这包括了两个层次，一个是外表与都市环境的关系，表示外表如何受外部环境影响，而产生一个对应都市环境的外观；另一层次关系，是来自外表与量体之间，两者之间因相对力量作用而产生中介空间，使得都市公园与建筑空间、外部与内部产生更强的互动性。

[3]楼板：楼板的产生来自于动线轨迹的延伸，板的剖面沿着轨迹拉伸。操作的因素包括了宽度、角度、厚度、弧度、参数的定义，形成空间连接的方式。

[4]结构：在建筑量体外表与楼板形成后，于建筑物相等的跨距做16组剖面，在剖面外框轮廓分别定义其结构，并考虑结构如何去支撑楼板量体与表面。当16组结构产生后，于每两组结构间做横向的支撑，形成建筑主要的结构体

↑Overall view of cablecar station／基地俯瞰图 ↓Traces & Fields／轨迹与场域

↓ 24hr density analysis ／ 24 小时密度分析　　Elevation view of dynamic study in the volumes ／ 量体动态分析立面序列图 ↓

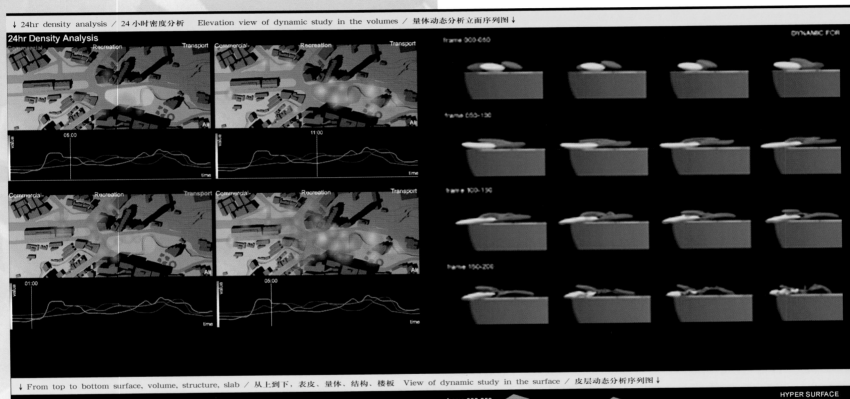

↓ From top to bottom surface, volume, structure, slab ／ 从上到下，表皮、量体、结构、楼板　　View of dynamic study in the surface ／ 皮层动态分析序列图 ↓

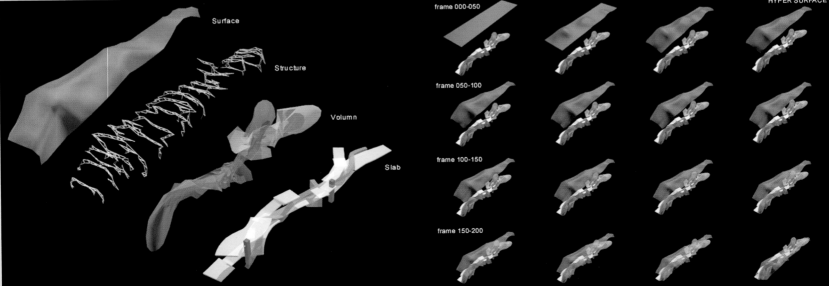

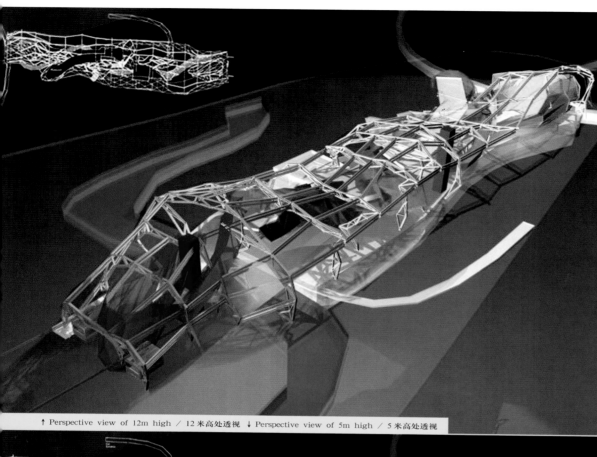

↑ Perspective view of 12m high ／ 12 米高处透视 ↓ Perspective view of 5m high ／ 5 米高处透视

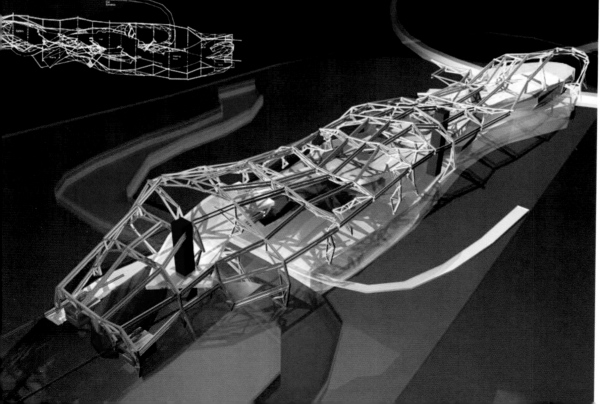

Architectural Typology / 建筑类型学

DESIGNER: Michel Hsiung

Thesis: The Standard Architectural Systems Can Create New Typologies When They Meet and Integrate. The Interaction of The Architectural System Acts as The Generator of The Network System, a Simple System That Can Evolve to a Complex One. My Investigation Ranges from Pipes, Slab, Truss and Skins.

The methodology of my approach was derived from the rhizome. The intricacy of looking for the unused has derived my investigation towards new typologies. Starting from the most generic forms, the possibilities of inhabiting the interstitial spaces that exist in most forms expanded my imagination. The question is not how to occupy the space but to fill it with organizational constrains that each space will lead the creation of the adjacent one. It is about how each component is related to the other versus a centralize idea that controls the whole. The decentralized approach of the network system allowed each move to inform the next. The formal aspect of the project is highly related to its function and relation of each component to the other. The digital media allowed me to do such an exploration taking out the constraints of conventional approach. The analysis went furtherer by integrating ideas from organic forms with architectonic constrains.

To enhance my theory I chose a random site as a continuum of my thesis. The site is an abandon steam power plant in Philadelphia. It is at the northern edge of the city where it was cut off from the rest of the city after the construction of a highway. Taken the same methodology, my intervention has taken the site and the building into a series of analysis that created a new systematic patterned to the city. Taken from the same language that was used in the prior investigation, the building emerged as part of the site providing occupyable spaces. The program of the project is for an E-commerce office where the business growth is related to its network expansion. The continuous flow that exists on the site will allow such expansion without guidance.

主题：标准的建筑系统可合并成为新的建筑类型，建筑系统间的相互影响，产生了可从简易转化成为复杂的网络系统。我的研究涵盖导管、楼板、桁架以及外壳。

我的研究方法取自植物根茎，由其错综复杂性引导出新的型态。从普通的形状开始，我的想像自多数形体可能存有的空隙；问题不在于如何占有空间，而是将它有条理的填满，让每个空间营造出毗邻的空间。本方案便是有关每个元素间的相互作用以及一个控制整体的中心思想，网络系统的分散让每个动作牵引下一个动作。本方案与其机能以及每项元素的关联有着密切的关系，数码媒体让我得以摆脱传统方式的限制而发展，更深入地分析有机类型引发的想法及建筑的限制这两者的结合。

为了加强我的理论，我随意选择一基地作为主题的延续。基地是费城一座荒废的蒸气发电厂；位于城市北缘，被高速公路与城市其他部分隔开。我使用同样的方式切入，将基地与建筑物带入一连串分析，创造一个新的城市系统组织。使用与前项调查相同的语汇，建筑物以部分基地的姿态出现，提供可复制的空间。本方案为电子商务办公室，其业务的成长与它的网络拓展相关，基地连续的流动性使其无需引导就可开展。[翻译 王俐文]

AFFILIATION: TEN W Architect
COUNTRY: U.S.A. / 美国

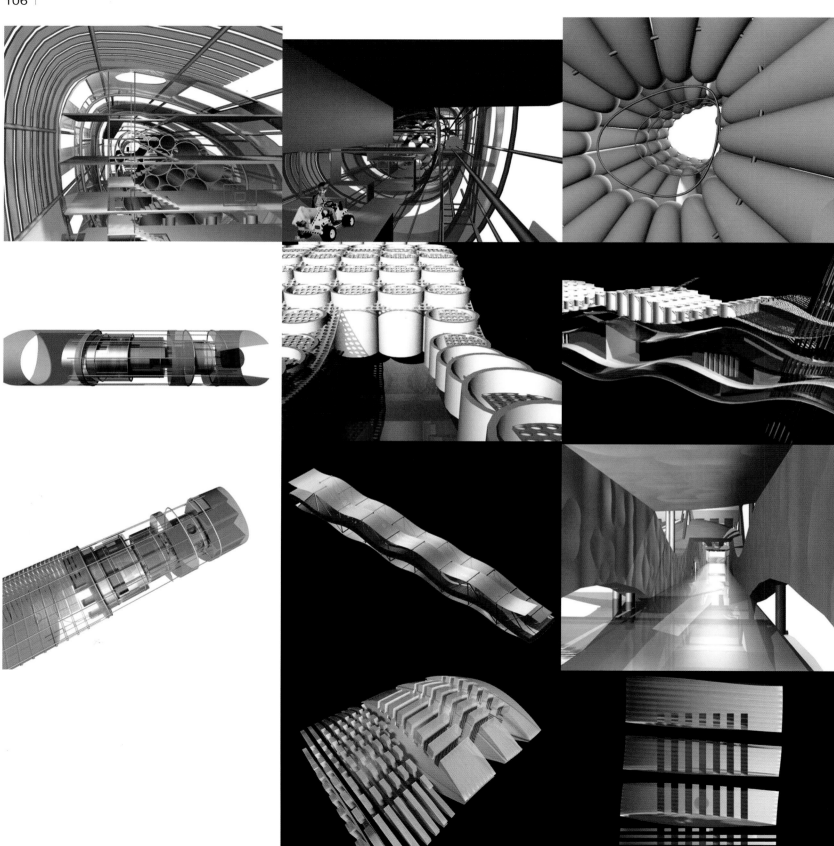

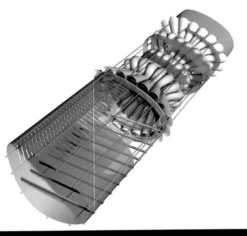
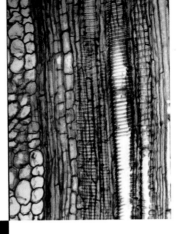

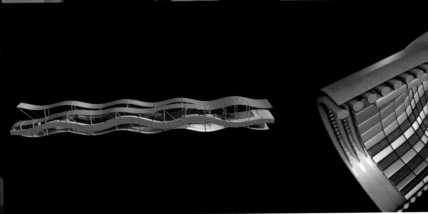

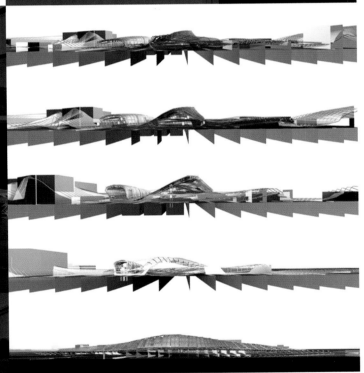

Music in Architect / 建筑中的音乐
DESIGNER: John Barnett

The following was originally created as a fifth year undergraduate thesis project for Ball State University (Muncie, Indiana, USA) in the spring of 2000. A 15 minute video animation was created presenting my ideas of representing sound and music visually / graphically. As seen in the following pages, these ideas are displayed in a variety of ways. Some are presented in the form of 'building and space' while others simply represent music and sound through the use of solid geometry. Music entices a broad spectrum and requires a vast array of representational devices. In any case, movement plays an important role.

Another crucial item to the thesis investigation was the definition of "architecture." Rather than investigating architecture directly, I chose to explore several of its components such as geometry, light / absence of light, movement, scale, etc... After these components were developed to music, they were then applied and implemented as design components. By first developing these components rather than the "architecture," a new architectural vocabulary was created upon their application, and thus, formed a more accurate translation and architecture unique to sound and music.

The creative limitation of architecture is the presence of rules: programmatic requirements, structural requirements, and even the presence of gravity. It is this reason that many other arts have to ability to be more expressive. For example, no rules exist in the world of art and music. Therefore, to accurately translate music in to a visual form, traditional "rules of architecture" were eliminated. The result is a more conceptual / psychological / visionary approach to architecture and solid geometry, where many of the ideas could only be constructed in a virtual world simply because of the above stated limitations and the fact that current architectural construction methods and contemporary geometries can not accurately show the graphic value of music.

Movement One: Sound = Geometry   My initial studies began with simple, direct, one-to-one conversions of sound to visual form. The geometry was derived from many different elements of sound, including the tonal quality, duration, articulation, intensity, and psychological response. The actual pitch of each note determined its placement in the z-plane. For example, a note low in pitch, with a heavy accent, may be represented as a visually heavy rectilinear volume placed low on the z-plane, while a bright, 16th note passage played by upper woodwinds, would translate into much lighter, smaller curvilinear geometries. Fig 1 represents the whales, which are crescendo half notes (gold geometries) with accents on the release (blue geometries).

Movement Two: Story-Telling   Differing from Sound = Geometry, Movement Two focuses on the concept of a musical work, and the emotions invoked, rather than creating new geometric tools from which to design. I began this process by selecting, "The Death Tree," composed by David Holsinger, and reviewing the composer notes. Then, after also studying the emotional quality of the piece, derived a list of criteria with which to work: estrangement, nonconforming, technology, disorientation, desertion, dominion, regression, isolation, host organism, might force, etc...

Movement Three: Urban Composition   The third and final movement represents yet another design approach, the structured method, as well as the culmination of all design tactics up to this point. The movement is designed around Eric Whitacre's, "Ghost Train," which has been translated at an urban level as a machine city. This city is composed of event nodes, each of which contains a different architectural theme, such as Light as Movement, Assemblage Architecture, Architecture as a Machine, and the Color Dream Sequence. Each of these event nodes are placed and sized according to the outcome of the matrix created by using the structured method. Movement three begins in one of several event nodes.

Within another such event node, The Overtones Sequence, the small linear geometries placed in the foreground represent the Sound = Geometry of the actual musical passage, while the large, transparent planes in the background represent the sounding overtones. During the last sequence, the viewer is exposed to the cityscape for the first time. Ideally, the city would be composed of many different musical passages, but because of time scheduling, is composed here of the same passage copied multiple times.

AFFILIATION: Ball State University
COUNTRY: U.S.A. / 美国

以下原是2000年春季Ball State University (Muncie，印第安纳，美国)大学部五年级的毕业设计方案。15分钟的动画影片作品体现了将声音与音乐视觉化的想法。这些想法以不同的方式呈现。有些以"建筑与空间"的方式呈现，有些以纯粹几何的方式呈现。音乐诱发宽广的频谱并且需要大量的设备。不论情况如何，运动扮演重要的角色。

另一个关键的项目是对于建筑定义的探讨。不直接研究建筑，反而探索它的几种组成，比如几何学、光与阴影、运动与尺度等等。这些组成发展成音乐后，就被用作设计的构成。首先发展这些组成而不发展建筑，运用它们时创造出了新建筑语汇，这样就形成一个更精确的转化和方法与音乐相匹配的建筑。

规则的存在限制了建筑的创意，例如：计划需求、结构需求甚至是地心引力。正是这个原因使得其他的艺术有较多表达能力。例如在艺术与音乐的世界里不存在规则。所以为了准确地把音乐转化为视觉形体，应将传统"建筑规则"排除在外。对于建筑和纯粹几何来说，需要通过更加概念化的、心理的和视觉的方式来表现，由于上述的限制以及如今建筑的工法和现代几何的存在使许多想法不能精准地体现音乐的价值，而只能在虚拟的世界里建构。

动作1 声音＝几何   我最初研究是从简单直接、一对一的声音与视觉形体转换开始的。几何形体由声音的许多元素中提取，包括音调、延续性、清晰度、强度与心理上的反

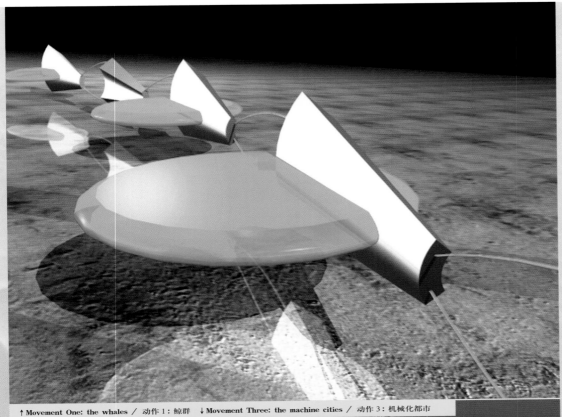

↑ Movement One: the whales / 动作 1：鲸群　↓ Movement Three: the machine cities / 动作 3：机械化都市

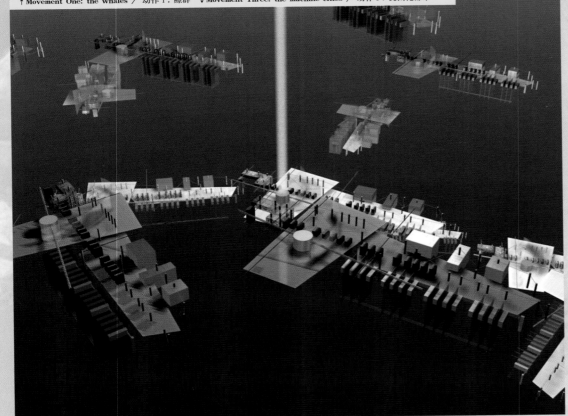

应。每个音符实际上的音高决定了它在 Z 平面上的位置。例如低沉的重音可置于 z 平面的低处，视觉上则产生沉重的线性量体；而明亮的木管乐音可转化成较轻盈、小的曲线几何。动作 1 代表逐次加强半音（黄金比）的 the whales，强调了舒放。

动作 2　故事叙述　不同于"声音＝几何"，动作 2 专注于一个音乐作品的概念以及所引发的情绪，而并不侧重于创造设计的新几何工具。我从选择 David Holsinger 所作的《死亡之树》开始入手，并回顾作曲笔记。然后再研究此曲情感上的特质，选取一份工作准则，例如：疏远、规章、科技、迷失、离弃、管辖、逆行、隔离、寄生母体、强权力量等等。

动作 3　都市作品　最后一个动作代表另一个设计意图，组织化的方法和到目前为止所有设计策略的高潮。它围绕 Eric Whitacre 的《鬼魂列车》而设计，在某种程度上它被转化成机械化的都市。此都市由事件节点组成，每个都包含不同建筑主题，比如光的运动、集合建筑、建筑机械、彩色梦的连续。每个事件节点组织系统计算出来的矩阵结果决定其位置与尺寸。运动 3 始于某个事件节点。

在这样的事件节点上，泛音序列，小的线性几何体置于代表"声音＝几何"的音乐章节前景下，而大的透明背景里的平面代表环境泛音。在最后的序列里，观赏者第一次暴露在都市的景观中。理想中，都市将由许多不同音乐章节构成，但因为时间表，它由一些经过多次复制的相同章节构成。[翻译 蔡咏岚]

↑ Movement Three: opening event node / 动作 3：开启事件节点   →Movement Three: the overtones sequence / 动作 3：泛音序列

A Building Which Exhibits Itself / 自我展示的建筑

DESIGNERS: Huang Ming Chen

Architectural Program   The building and its site are exhibits of themselves. The deployments of all architectural elements that define the building, which exhibits itself, are exhibits The building will have exhibits about the process of making itself through the exploitation of design technologies. The exhibit consists of drawings and models depicting the process that exhibits the architecture. The building, which exhibits itself, is a prototype demonstrating variations of the formal system; the transformations and generations of each variation; and the presentation of various transformational processes.

建筑计划   建筑物与基地是自身的展示物。所有界定此建筑物的建筑元素的部署本身都是展示。通过设计技术开发的建造过程将在以建筑中得到展示。展览包括描绘建造过程的图画与模型。它是示范正规系统的差异和每个差异的变化与衍生以及不同转化过程呈现的典型。[翻译 蔡咏岚]

↓ Analytical—Conceptual Extraction / 分析—概念摘录

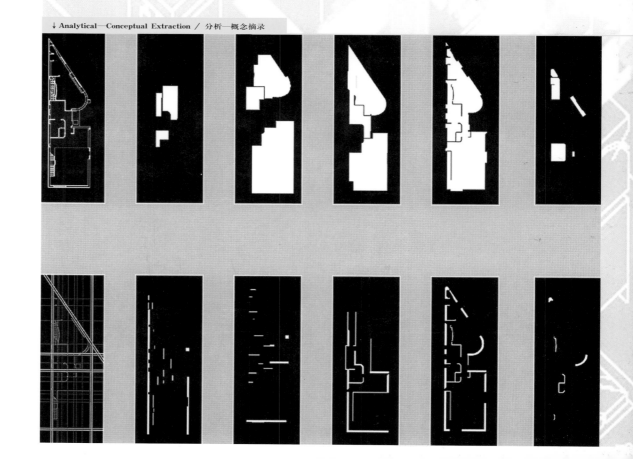

COUNTRY: U.S.A. / 美国

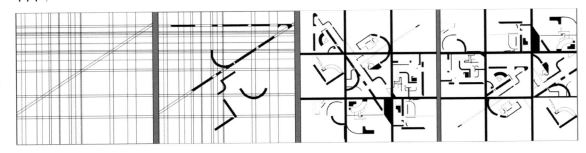

←Fragment Recomposed ／ 改编的片段

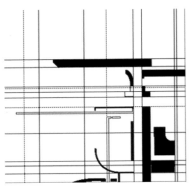
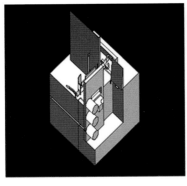
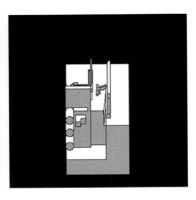

←Relief Study ／ 对比研究

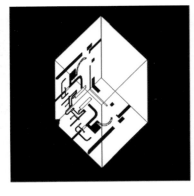
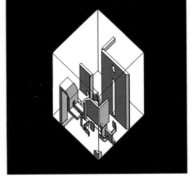
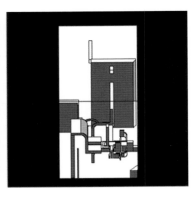

←Interlaced Interpretation ／ 交织的阐述

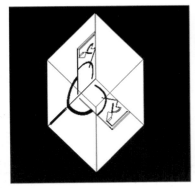
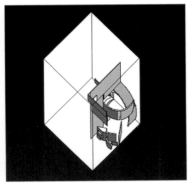
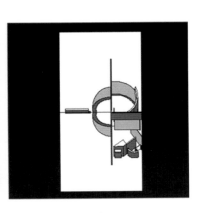

←Interlaced Interpretation ／ 交织的阐述

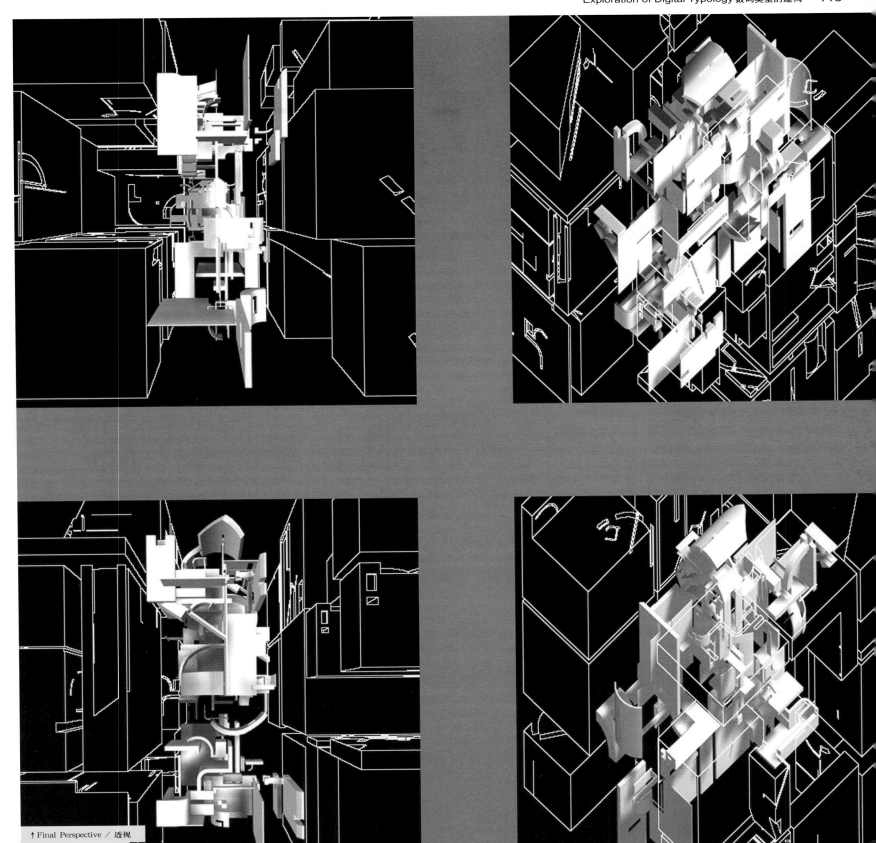

↑ Final Perspective ／ 透视

Just-in-Time Infrastructural Prototype — How to Organize Space from Time Instead of How to Organize Time from Space

及时型的公共设施原型如何以时间组织空间，替代以空间组织时间？

DESIGNERS: Igor Kebel

In a word   Just-in-Time Infrastructure investigates a possibility for conceptualizing urban space from networking time. Its policy is structured as a materialized infrastructural joint for the specific European market, based on service-economy. The vitality of this prototype development is in the spatial distribution of a program derived from the time. The programmatic distribution is thought throughout the re-organization of daily living and working protocols, telecommunication transfers, and spatial progression of flow stimulants.

The 24/7 Lifestyle   One impact of the ultra-liberal new economy, driven by the 24/7 motivation of service-based businesses, is the constant implantation of fresh organizational principles into our societies, where work continues anytime, anywhere, and where private- or state-, political- or cultural institutions share the same infrastructural interests. This merciless race of information flows on a shared infrastructural platform produces elementary changes in modern lifestyles; the traditional distinction between ideal forms of living and working has become almost meaningless. With regard to these new lifestyles, how can architecture act reflexively in its day-to-day practices? Even more importantly, what are the demands of these new users (clients)? What kind of product do they need? Which new organizational clusters can guide (rather than follow) these economic and social imperatives?

City inhabitants, city users, commuters, and metro-businessmen are, according to Guido Martinotti, clustering a pattern of modern metropolitan areas. Yet their range of mobility, in a classical sense, is determined by their territorial enclosures. Through the ICTs (information and communication technologies) supplementation of the transportation of services and tasks, parallel networks have been launched. Increased mobility and the duplication of infrastructures are jamming daily working and living routines. Is there a way to optimize of these complex infrastructural links and knots of simultaneity?

Shared Infrastructures   This project interprets the principles of Just-in-Time manufacturing the on-demand production method invented in the 1970s by Ohno Taiichi of Toyota in architectural and infrastructural terms, and applies them to four pilot participants: The City of Amsterdam as a representative of public interests, and three private companies that operate in a service-based mode a cable services provider (A2000), a newspaper (Het Financieele Dagblad) and a "smart-car" prototype developer (Mercedes-Chrysler).

These companies share similar service- and infrastructural demands for their business adaptation and proliferation. They all supplement their classical products with the mediated services that need to be active 24 hours a day. To say conclusively and in a manifestable manner: From clock-time to flex-time towards just-in-time. From political economy to new economy towards perfect economy. In architectural and urban terms, this project proposes organizational solutions that instrumentalize the simultaneity of networking events and places on a selected location in Amsterdam.

4 into 1   Four programmatic entities domestic, mobile, public, and work crucial ingredients of the Just-in-Time Infrastructure, are merged into seamless one. This does not eliminate distinctions between distinguished spatial enclosures; it simply creates shared and flexible spaces among all the programs included. Such a management of users is clustered in a 24-hour timeline in order to operate without delays and in all situations. Residents, users, commuters, metropolitan nomads and telecommuters are served in real-time situations. They work and live anywhere and anytime.

简而言之   及时型公共设施探讨了从网络时间将都市空间概念化的可能性。它的目的是成为物质化基础结构的接点，从而服务于特定的欧洲市场。此一原型发展的活力在于由时间衍生出计划的空间配置。系统化的分配通过日常生活、工作礼仪、通讯传输、流动刺激的空间发展等等的重新建构而建立。

24/7 的生活方式   极端自由的新经济结构的冲击之一，是新组织规则不断地纳入我们所属的社会。在这个社会中，在任何时间、地点、私密或公共、以及任何享有政治文化的机构中的工作都持续不断。无情的信息竞赛在一个共享的基础平台上流溢，它对我们的现代生活产生树根本性的改变。传统对于理想的生活与工作几乎都没有意义了。关于这些新的生活方式，建筑对于这些日复一日的实践如何反应？更重要的，这些新客户的需求是什么？他们需要什么样的产品？什么样的新组织将引导（非跟随）这些经济与社会责任？

据Guido Martinotti所言，都市居民、使用者、通勤者、都会商人在现代都会区域里集中，形成一种模式。但他们机动性所及的范围，从传统角度去判断，由自身领域的范围界定。通过服务以及作业传输的信息与通讯科技的供给，平行的网络已经建构起来。机动性的增加与公共建设的倍增干扰了日常工作与生活常规。有将这些复杂的结构连接起来解决所发生的难题的办法吗？

AFFILIATION: ELASTIK
COUNTRY: Slovenia / 斯洛文尼亚

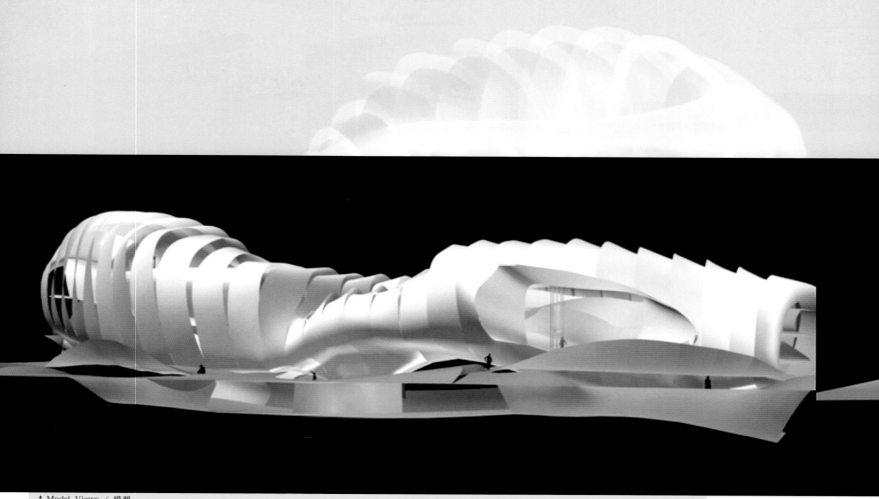

↑ Model Views ／ 模型
↓ Network of Programs, What is normally practiced in our daily routines in dislocated environments is within this prototype tested at a single location ／ 计划的网络，我们平日在孤立环境中所依循的常规在此进行测试

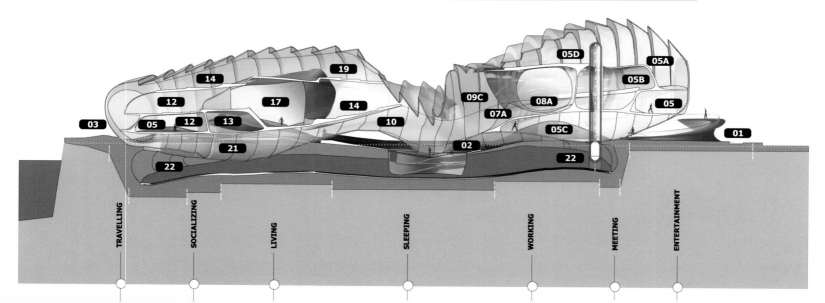

TRAVELLING SOCIALIZING LIVING SLEEPING WORKING MEETING ENTERTAINMENT

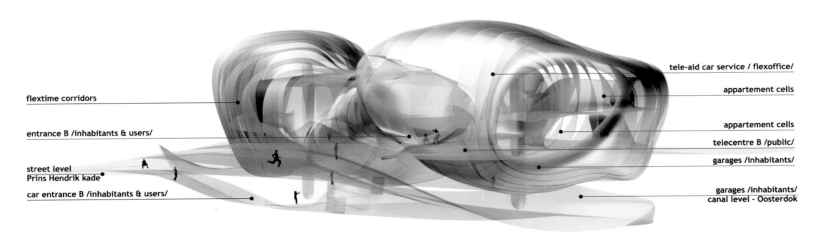

tele-aid car service / flexoffice/

appartement cells

flextime corridors

appartement cells

entrance B /inhabitants & users/

telecentre B /public/

garages /inhabitants/

street level
Prins Hendrik kade

car entrance B /inhabitants & users/

garages /inhabitants/
canal level - Oosterdok

↑ Integral Prototype, The programming of a prototype without hierarchies: infrastructure and content are inseparable. 4500 ㎡ with the vertical, horizontal and diagonal connectivity / 整体原型，一个没有阶级原型的规划：基础设施与内涵是不可分割的。4500 ㎡ 有垂直水平与对角的连线

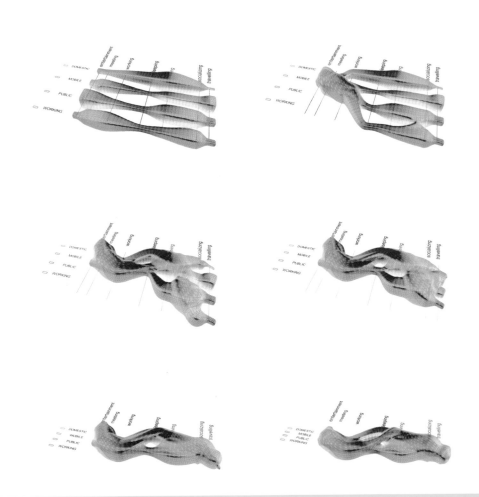

共用的公共设施　这个案例诠释了恰当时间的原则，基于要求的制造方法，就建筑与工业而言，作用于四个参与者:阿姆斯特丹市作为公共利益的代表，三个以服务为基础的私人公司有线电视服务业者(A2000)，报社(Het Financieele Dagblad)与"smart-car"(奔驰车款)原型的开发者梅塞德斯·克莱斯勒汽车公司(Mercedes-Chrysler)。

这些公司共享相同的服务与基础设施需求。它们都以需要24小时运转的调配服务来支持它们基本的产品。最后总结并强调：从按时到弹性时数到恰到时间。从政治经济结构到新经济体制再到完美的经济体制。就建筑与都市而言，此案例提议组织性的方案，这种方案促成阿姆斯特丹所选的网络活动与场所的刺激。

四合一——四个计划性的实体家庭的、活动的、公共的和工作的决定性的要素合并为一体。这并不会排除特定空间之间的区别；它只不过创造共享的弹性空间。这样对使用者的管理被集中在一个24小时的时限上以便在任何时刻都可以准时地操作。居民、使用者、通勤者、都会游牧者与在家办公者可以享受及时连线服务。他们可以在任何时间、任何地方工作。

↑ Four into One, Animated geometrical calculus, which supports initial concept of four domains (domestic, working, mobile and public) merged into one / 四合一，支持最初四个领域(家庭的、活动的、公共的和工作的)概念的模拟几何计算合而为一

A Terminal at the Haifa Port / Haifa 港的转运站

DESIGNER: Ori Scialom

This project is trying to use computer analysis power into the planning process. The generation of "forms" and program relations are based on computer "analysis" of "data" related to the site and program. The "analysis" is based on computer simulations using animation. Today, as result of wide process in digital tools, we are able to collect and analyze a wide spectrum of data. This data was less accessible and almost useless to planners a few years ago.

Architecture today is facing new challenges. What are the relations between "form" and Information? Can "data" be a "form" generator? Can we produce a planning process based on information? Do we need all the information that floods us? This project is trying to deal with some of thesechallenges.

The Program   The building install boats and train terminals, passengers facilities like hotels, shopping areas and cultural activities. It also has a direct car access to the highway.

The place   The Terminal is located at the port of Haifa, Israel. This location is a hub to Boats, Trains, buses and cars. The amount of passengers is changing constantly, characterized by peaks of arrivals and departures. This situation seemed suitable to deal with computer analysis.

The Technique   Data regarding the quantities in relation to time was collected. (Based on research done at the Technion Institute, Haifa). This data was put into diagrams showing its changes during a day and week cycles.

The Simulations   The "data" was put into simulation models, giving a wide spectrum of changes. The result is a dynamic model simulating "data" that can be recalculated and examined at each frame of animation. Three types of simulation were made:
— Quantities of users during time cycles. This phase is major in understanding the form volumes and configuration.
— Simulation of the active areas in the structure during time cycles. The phase helps in performing program layout inside the building and the gathering of functions according to time-working zones.
— Simulation of predefined parameters. For example, defining a parameter for the connection sea-city produced different values on winter and summer. These values helped in understanding the sort of connection needed at the site.

此案例尝试把电脑分析的力量用于规划过程。"形"的衍生与计划关系基于电脑与基地和计划相关的"数据 、"分析"。分析基于运用动画的电脑模拟。在今日，数码工具被广泛使用，我们可以搜集分析广泛范围的数据资料。这些资料在几年前，对于规划者而言几乎很难接触不到并且毫无用处。

今日的建筑面临新的挑战。形体与信息之间的联系是怎么样的？资料可以作为形体衍生的工具吗？我们可以制造一个基于信息的规划程序吗？我们需要所有淹没我们的信息吗？此案例试图处理这些挑战。

计划   此建筑内有船，铁路车站，诸如旅馆的旅客便利设施，购物区和文化活动，也有高速公路直接的交通连接。

场所   车站位于以色列海法港(Haifa)，是船、火车、车辆的活动中心。载客的数量不断变动，由到达与离开的尖峰描绘出特性。这样的情况似乎适合用电脑分析处理。

技巧   搜集了关于时间的数据。(基于在 Haifa 技术机构所做的研究)将资料绘成图表，显示了日或周循环的变化。

模拟   资料输入模拟模型，并可在很大范围内变化。该成果是个动态的模型，模拟了可以在各个动画镜头中被重新计算并检视的数据。创造的三种模拟分别是：

AFFILIATION: Ori Scialom Architect
COUNTRY: Israel / 以色列

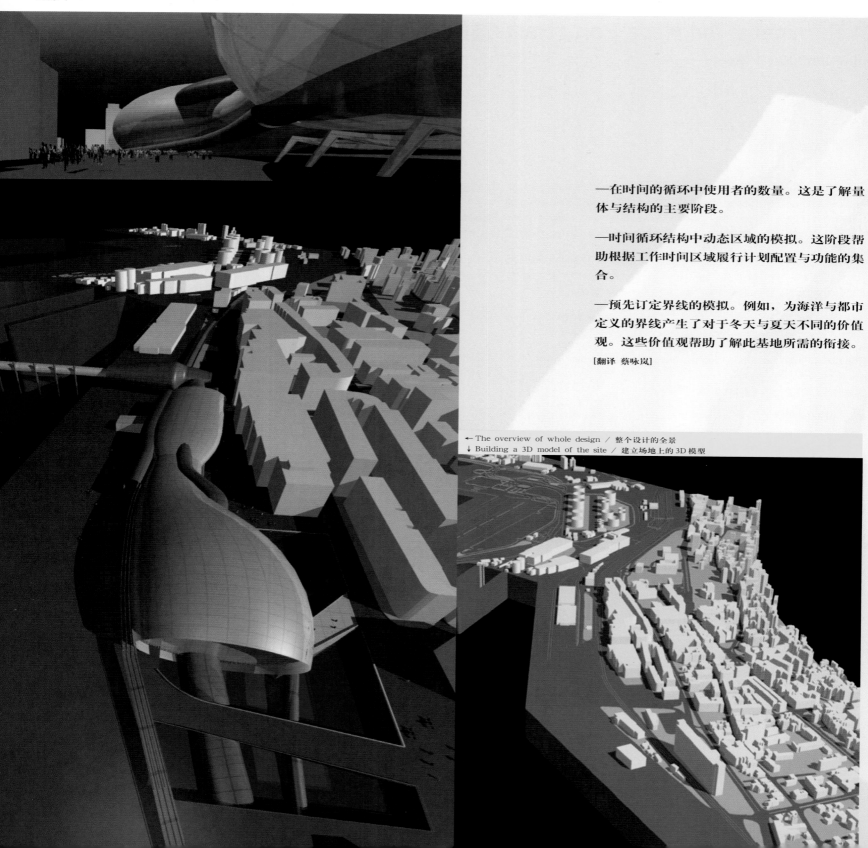

—在时间的循环中使用者的数量。这是了解量体与结构的主要阶段。

—时间循环结构中动态区域的模拟。这阶段帮助根据工作时间区域履行计划配置与功能的集合。

—预先订定界线的模拟。例如，为海洋与都市定义的界线产生了对于冬天与夏天不同的价值观。这些价值观帮助了解此基地所需的衔接。

[翻译 蔡咏岚]

← The overview of whole design ／ 整个设计的全景
↓ Building a 3D model of the site ／ 建立场地上的 3D 模型

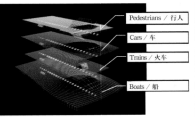

Pedestrians / 行人
Cars / 车
Trains / 火车
Boats / 船

The Simulation Models / 模拟模型

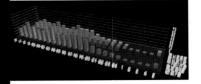

Sunday to Thursday / 周日到周四

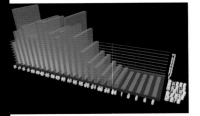

Accumulative Middle Week / 累积的半周

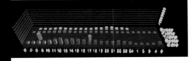

Weekends / 周末

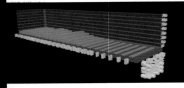

Accumulative Weekends / 累积的周末

←Traffic Flux Data / 交通通量数据

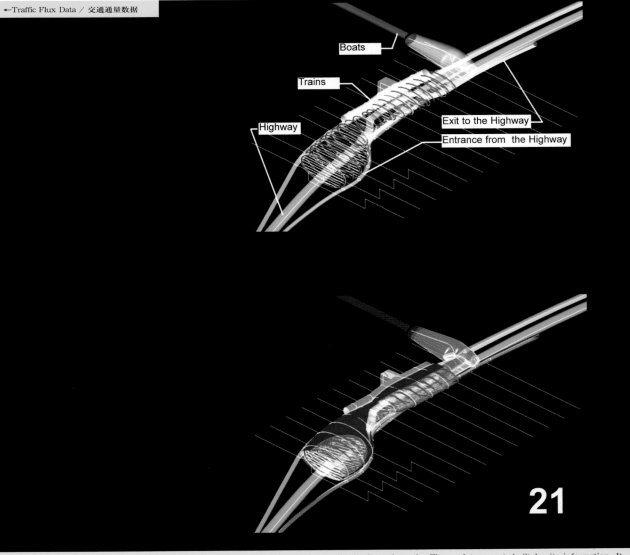

Boats
Trains
Highway
Exit to the Highway
Entrance from the Highway

21

↑ Since the design study was built mostly on the traffic imformation. The complex is situated on the Flows of transport, built by its information. It gathers them together and makes the connections between them, and their passengers / 设计的研究大都基于交通信息。复合体置于运输流之中。它搜集信息并以之与乘客沟通。
↓ The peaks'model have been processed and smoothed. This processing generated an array of veins for the traffic flow to and through the site / 模型经过处理与平滑。此过程产生穿越基地的交通脉络

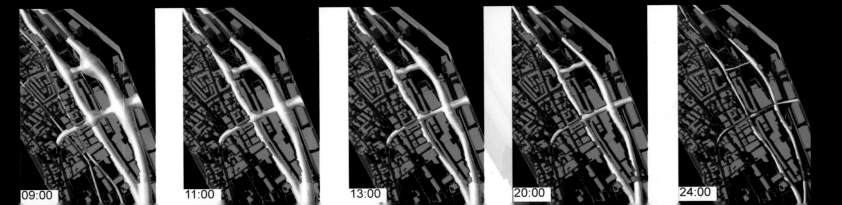

09:00　11:00　13:00　20:00　24:00

Dynamic Linear Membrane Infrastructure For Festival /
节庆的动态线性薄膜公共设施

DESIGNER: Shih Weng / 翁狮

A dynamic linear membrane infrastructure in the central district of Taipei is an infrastructure for festival. The new skin attached on the existing building is shaped by the skeleton (sliding ceiling, folding platform...) and the muscle (capsule cinema).

The linear membrane as a medium of time.
The linear membrane is a performing stage of image shadow and light.
The linear membrane is not only a container of information. It become information itself.
The information becomes an architectural material.

The experience of walking through the street changes when the skin changes. The skin changes when the skeleton changes, when the muscle changes, when the image changes, when the activity changes, when the time is running.

位于中国台北市中心的一个动态的线性薄膜公共设施是举办节庆活动的场所。安装在现存建筑物上的新表面由骨架(滑动天花板、折叠平台)和肌肉(太空舱电影院)组成。

线性的薄膜是时间的介质。
它又是光与影的表演台。
不只是信息的容器。它本身就是信息。
信息就是建筑材料。

穿过街道的体验随表面变化而变化。表面随骨架变化而变化，当肌肉改变，影像改变，活动改变，时间流动，表面也随着变化。[翻译 蔡咏岚]

AFFILIATION: 1999-2000 MArch The Bartlett School of Architecture
COUNTRY: Taiwan, China / 中国台湾

↓ Axial Elevation ／ 轴线方向剖面

Section D

Time : 2001 Feb Taipei Film Fe

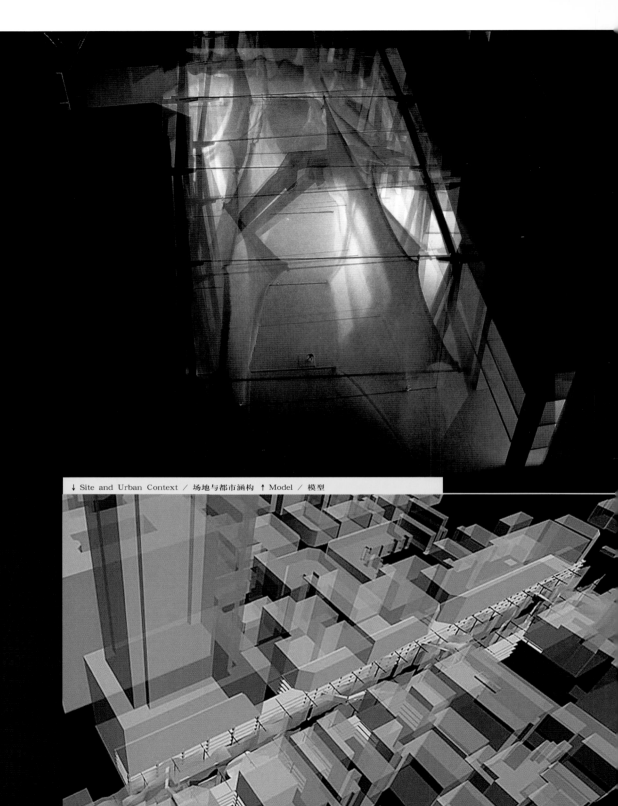

↓ Site and Urban Context ／ 场地与都市涵构　↑ Model ／ 模型

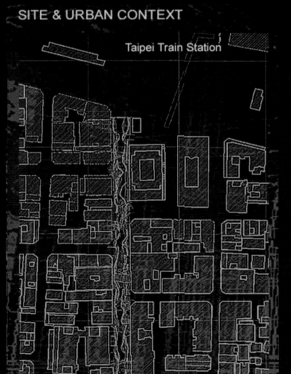

SITE & URBAN CONTEXT

Taipei Train Station

Scale-less / 无尺度

**DESIGNER: Bounding Box / 界域箱子 Canan Aka Y.T.U; Circlelink / 环状连接 Nuket Zeydanli M.S.U;
The Archetype / 原型   Oytun Aren I.T.U; Immanent Viewpoint / 内视观点 Koksal Aksoy O.G.U;
Point Hunt / 猎取点   Bulent Onur Toraman Y.T.U; Space Boost / 空间升举 Tansel Dalgali Y.T.U**

The project is a statement that is communicated through the work of a studio: 8 participants and 2 conductors. The initiative idea for the studio had been formulated as follows:

While computers have replaced the ruler and the t square in architectural offices, they have not truly transformed the habitual thinking with that equipment yet, even though we are pretty much aware of the fact that to start with a cardboard model will end up with a cardboard architecture. We still work with hierarchy of scales- first master plan, then incrementally zoom for details: the master plan must maintain its integrity throughout the detailing process. Against such a boring linear process and obsession with origin, we remind something everybody knows: working with computers, there is no scale -any scale at any time-, and no origin no original master plan in the drawer but a folder in the PC that we continuously work on-. In other words, as it also makes it so easy to edit, computer is an agent for endless manipulations. Sit back and enjoy a line that goes to infinity, or a light that does not cast shadows. The project is the (architectural) object of weekly manipulations in the tabula rasa of screen space, the context is the first line or particle thrown into the space that would disappear in the following weeks, the program is (eyes) view, the end is the end of the semester, the hope is liberation through submission.

As in most of the successful studio works, we observed deviations from the initial plan through the course of the exploration. They appeared in the form of shifts from the actual object to the idea of the object in the manipulation process. We also challenged the idea of origin in manipulative menus of the software referring to certain natural phenomenon giving acceleration to particles: raindrops falling on the ground towards self referential operations just by twisting their initial logic and using them in combinations. It became like a childs game where there are no predetermined rules or images, no end; nobody wins. Eventually, we realized that it took discipline to break free from the resistance of submitting oneself to what he was working with, to let it take him wherever it might have lead. It was liberation.

本案例是一个八位成员与两位指导老师组成的工作室的声明。工作室所制定出的初步的想法如下:

虽然电脑取代了建筑工作室里的丁字尺与三角板，就算我们很清楚从纸板模型出发的设计会做出那样建筑，它也并未全然改变使用该工具的习惯性想法。我们还是以等级推演的尺度操作为先，先处理大配置，然后逐步地拉到细部上面:主配置必须贯彻到整个细部设计中。对于这么一个无聊的线性操作和出于对起始概念的着迷，我们提醒一个大家都知道的事情:使用电脑在任何时候都没有任何的局限，没有源头，没有最初的计划，只有电脑里一个我们持续作业的资料夹。换言之，在容易编辑的同时，电脑是无尽操作的中介。坐下来享受一条无限长的线或是一盏不投阴影的灯吧。本方案是每周一次的在屏幕白板上的操作，其涵构是第一条线或说是在即将来临的数周内被丢入空间中消失的微粒，本提案的来源是眼睛的观点，它的结束是学期结束，所抱的希望是通过妥协而得到的解放。

如同多数成功的工作室作品，我们观察最初计划的偏差。它们以目标到实际的操作过程出现概念的转移为特点而出现。我们也挑战了一些谈论与特定自然现象相关的软件操作菜单的起始概念促使粒子加速:雨滴落到地上仅仅通过扭转它们初步的逻辑并且将之组合就朝向自我指示操作的方向。它逐渐成为儿童游戏般没有预定规则或影像，也没有终点，最后没有人获胜。渐渐地，我们领悟到要是不墨守成规有所创新，是需要自律的，这是一种解放。[翻译 蔡咏岚]

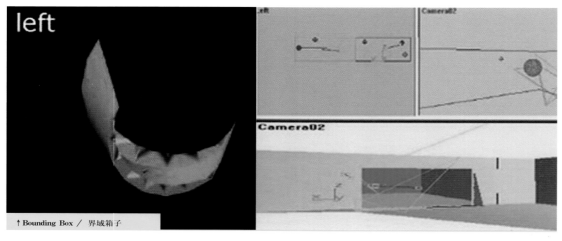

↑ Bounding Box / 界域箱子

AFFILIATION: Bom Design Group
Studio Coordinators: m kutuk MArch SCI-Arc, elifk MSc Columbia University
COUNTRY: Turkey / 土耳其

The Digital Tower / 数码塔楼

DESIGNER: Borislav Ignatov

The digital tower project is developed exclusively for the 2000 FEIDAD award contest and presents a simple idea. The idea of encoding essential digital information into an architectural tissue. The concept defines the term digital architecture as a system for incorporating information into an order of architectural elements. The very essence of the digital architecture idea is the substitution of any pair of architectural ordered elements for binary 0 and 1 logical units. The digital tower is a product of the digital architecture concept and is a project for a building in which the sequence of facade elements on each floor presents the floor number in binary. Then the entire tower elevation becomes a discrete code representation of the essential human knowledge and architectural context.

The creation of the digital tower evolves through the next three logical steps:

[1] Expression   Setting up the information to be encoded into the architectural object
[2] Translation   Transforming the information into universal binary code
[3] Conversion   Substituting architectural elements for the binary 0 and 1 in the building tissue

The digital architecture concept addresses the very basics of architecture. It creates a new kind of architectural order that does not restrict the design of specific architectural elements. A new feature of the design is that it contains and presents information that may be understood through reading the ordered elements. Instead of using extraneous tv screens and computer displays, the architecture itself becomes the display matrix. The information displayed may be as simple or as complex as the designer chooses to make it. The system may be used simply as a way to create a coherent and pleasing design or with the goal of imparting specific information. Digital architecture, as defined in this project, is a new tool for creative expression. The possibilities are infinite.

数码塔楼的方案是特地为 2000 FEIDAD 竞赛而设计的，它表达了一个简单的想法，就是将必要的数码信息编入建筑的组织中。这构想将数码建筑一词定义为体现信息具化为建筑元素的系统。数码建筑的基本构想是将任何一组建筑元素替换成二进制的逻辑单位。数码塔楼是数码建筑概念的产物，每座楼以二进制排序的立面元素代表各楼层层数。所以整个塔成为代表人类智慧精华与建筑涵构的不连续编码。

数码塔楼的创作包括以下步骤：
[1] 表情   建立要被编码为建筑元素的信息
[2] 转化   将信息转化为通用的二进制编码
[3] 换位   以建筑元素替代建筑物中的二进制编码

数码的概念是此建筑的基础思想。它提出了新的建筑准则，此准则不会限制特定建筑元素设计的。此设计的特点是，通过阅读顺序排列的元素可获得包含于其中的信息。在不用外来的电视屏幕和显示器的条件下，建筑本身反而成了显示的母体。显示的信息随设计者意图，可简单亦可复杂。这个系统不仅可以用于创造连贯合意的设计，也可以用于传播特定信息。此提案中所定义的数码建筑是创造力的新工具，其中蕴藏了无限可能。[翻译 蔡咏岚]

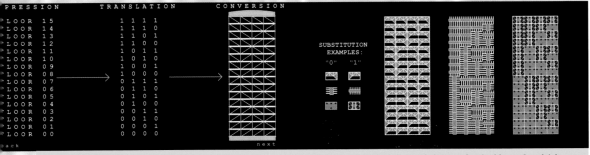

↑ Transforming the information into universal binary code / 将信息

↑ Substituting architectural elements for the binary 0 and 1 in the building tissue / 以建筑元素替代建筑物中的二进制编码

AFFILIATION: Arch-E-Structure Design Group
COUNTRY: U.S.A. / 美国

Neuronal Architecture Processor — How to Generate Architecture Directly out of Neuronal Spikes

神经元建筑处理器—如何从神经元突起中直接产生建筑

DESIGNER: Manfred Wolff-Plottegg, Wolfgang Maass / Programming: Harry Burgsteiner, Andreas Gruber

Processor 1 simulates 3 biological neurons. The "spike train" output is transmitted to Processor 2. A script, simulating a "DNA-transcription", transforms the incoming spike trains into data sets, binary strings. By autolisp the "genes" of the spike train ("DNA") are transformed into autocad commands. Thus the incoming spike-trains permanently generate new 3D-solids.

This installation is an experimental set-up to demonstrate a principle and to prove our thesis on digital creativity. The principle is the following equation: A sequence of pulses (spike trains) in biological organisms = binary streams of signs (bit strings) = data (coordinates, vectors) interpreted as geometric bodies (solids). This equation brings together three spheres of reality that are traditionally considered as distinct:
— the world in our head: information, imagination and ideas in the way they are actually coded in our brain
—the world of digital data processing, represented by binary streams of signs as communication vocabulary between digital processors
— the world of generating new images / spaces / architecture

These three spheres of reality are brought together under the common denominator of information processing. Thus the installation demonstrates that the generation of images / spaces / architecture must no longer be perceived as anthropocentric / expressionistic. Through the use of the computer it is possible to emancipate oneself from established patterns of behavior, evalutation and production. This has consequences for our understanding of art, architecture and human versus machine/digital creativity: it proves our thesis that creativity (in this case production of architectural images) is the result of a "neuronal machine" and as such can be released from the human brain.

处理器 1 模拟三个生物神经元。"神经元突起列串"被发送到处理器 2。模拟 DNA 样本的底稿将收到的神经元突起列串转化为二元的数据资料。通过一个自写程式，神经元突起列串的基因，也就是 DNA，被转成 AutoCAD 指令。如此输入的神经元突起列串持续产生新的 3D 实体。

这个装置示范了数码建筑的规则，并验证我们对数码创造力的论点。以下的方程式就是提出的原则：处于生物有机状态的脉动（神经元突起列串）= 二元信息的河流（记忆体线串）= 被视为几何体的数据（整合系统与导航器）。这个方程式统一了传统上被认为相异的三个领域：

我们心智里的世界：信息，想像力和思想在我们脑中实际上的编码

数码资料处理的世界：作为数码处理器间沟通语言的二元信息河流

产生新影像、空间或建筑的世界

这三个现实的领域通过信息处理的分母结合。此装置示范了影像、空间和建筑的产生不应该终止以人为中心的或者说表现主义的传统理解。透过电脑的使用，从现有的行为模式、价值观和创作中解放。我们对艺术、建筑以及人与机器或数码创造力的认识将有不同的结果：它证明了我们对创造力（指建筑影像的生产）是一个"神经元机器"的产物因此可从人脑中释放出来的论点。[翻译 蔡咏岚]

↑ Spike trains, bit strings, binary codes, DNA, genes, reading matrix, RNA ／ 神经元突起，记忆体线串，二元编码，DNA，基因，阅读母体，RNA

AFFILIATION: Interdisciplinary Team, University of Technology, Graz, Austria
COUNTRY: Austria / 奥地利

921 Seismic Preservation Museum — Visualizing the Earths Force/
921 地震保存博物馆—地球力量的形象化

DESIGNER: Chia-hsun Lee/ 李佳劻

Background   The earthquake occurred in Chi-chi on September 21, 1999. Its unpredictable power had made great losses to our people in Taiwan,China. After the great shock, this small island was changed its face. This chosen site that was used to be an elementary school and a junior high school shared the same field was separated into two parts by the fault line. The force of earth damaged both artifacts and nature, which only exist in our imagination. In this project, I try to figure out the force of earth through a digital approach.

Sequence of forming the invisible forces   [1] Land Forcethe cracking fault and the catastrophic school buildings create main elements in landscape. [2] Overturning Shakethe museum is lifted up and turned over as the ground surface elevated. [3]Tearing Forcethe torn-up form causes fractures in the joints of different components. [4] Twisting Quakethe metamorphic process is visualized through the roof and the interior of the museum. [5] Reborn Strengththe community center arises from the building wreck. [6] Collecting Memoryvisitors are invited to mourn for the victims, to share the memory of the locals, and finally to celebrate the rebirth of the region.

场景　　中国台湾集集地区在1999年9月21日发生了地震。

无法预测的力量造成中国台湾人民重大的损失。在巨大的震动后中国台湾地区的面貌整个改变了。选择的场地曾经是中小学用地，在地震中因断层线通过而一分为二。无法触及，损害了人造物体与自然。在此案例中，我试图通过数码的手法解析地震的力量。

隐形力量的组合序列　　[1] 地震力撕开的缝隙和灾变后的校舍是地景上的主要元素。[2] 剧烈的震动导致地表面升高，博物馆倒塌。[3] 地震造成的裂痕导致各部位接头的断裂。[4] 扭转的后果通过博物馆的屋顶和内部空间将变形的过程视觉化了。[5] 重生的力量—社区中心在建筑物的残骸中升起。[6] 大批访客被邀请来此纪念遇难者，并与当地居民共同的回忆往事，最后庆祝这个地区的重生。[翻译 蔡咏岚]

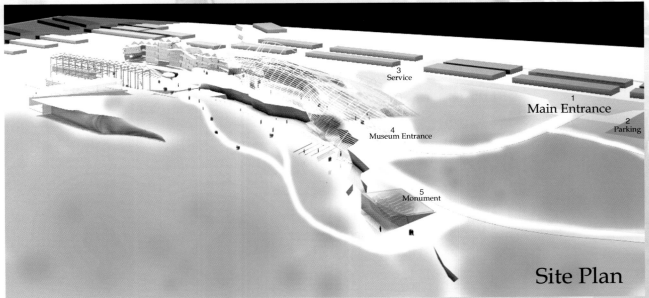

3 Service
1 Main Entrance
2 Parking
4 Museum Entrance
5 Monument

Early concept for site planning / 早期基地规划的概念
1. Main Entrance / 主入口
2. Parking / 停车
3. Service / 服务
4. Museum Entrance / 博物馆入口
5. Monument / 纪念碑

Site Plan

AFFILIATION: Graduate Student, Department of Architecture, National Cheng Kung University / 中国台湾成功大学建筑研究所研究生

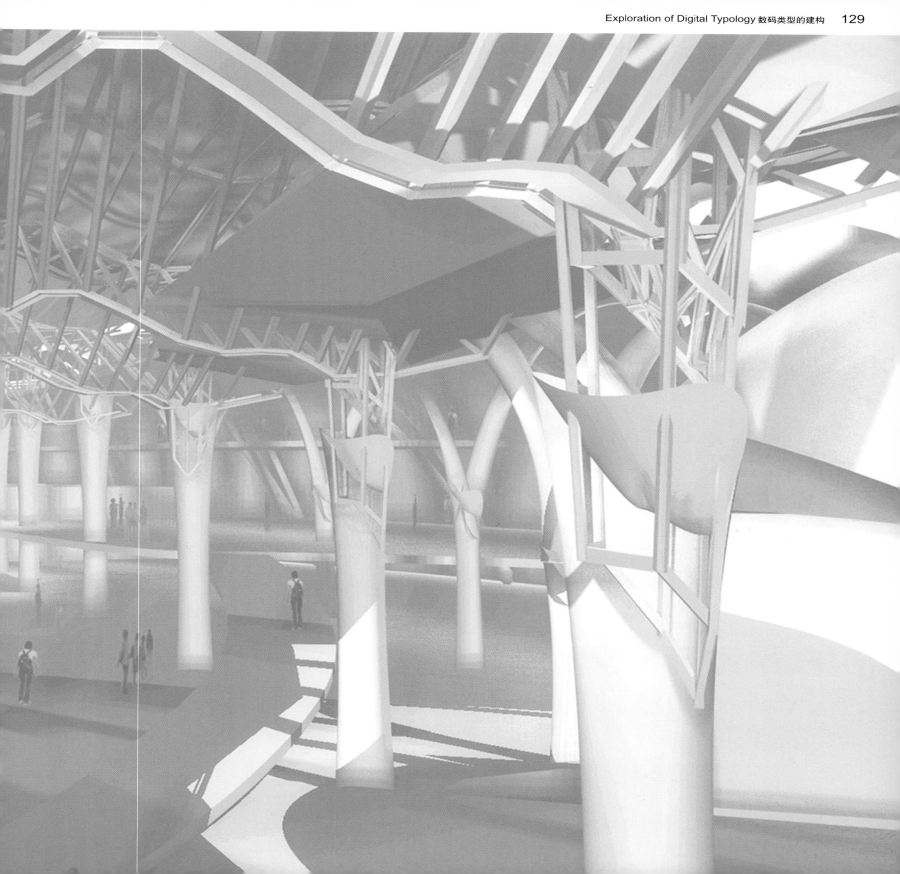

Evolving

Evolving Concept of Space

Humanity has always maintained a plural concept of space. For instance, there is ritual space as explained in anthropolgy, expression space as proposed by musicology, etc. All of these types of architecture's concept of space have their own unique characteristics. Because architecture must concern itself with the spatial functionality, architecture's concept of space is relatively simple and constructible. In the West, spatial concepts begin with mass space in ancient Egypt, then to the geometric space of the Greeks, mysterious Gothic space, the dynamic space from the Renaissance to the Baroque period, and finally modern and post-modern concept of space; while in the East, there is also the concept of void and solid space, again showing the evolutionary changes of our concept of space throughout history. Architecture often changes with the times. At the end of the 20th Century, humanity's concept of space has suddenly experienced a great change with the arrival of the internet. Cyberspace or networked space now exist alongside actual space. Computer simulation technology has allowed buildings to be "built" in virtual space that were never intended to actually be constructed. This new virtual space has not only influenced the direction of architectural design but has overturned our existing spatial theories. Now architectural space can be as multifaceted as other forms of space, including ritual space, textual space, etc.

Concept Of Space

空间概念的演化　　　　人类向来对空间具有多样的概念。人类学上的仪式空间、文学中的文本空间、以及音乐学所论及的表情空间，都具有十分广泛而且想像力十足的特征。建筑由于必须顾及功能，因此空间概念必须比较单纯而且易于建造。自古埃及的量体空间、古希腊的几何空间、哥德的神秘空间、后期文艺复兴到巴洛克的动态空间、现代与后现代空间等西方空间概念，以及东方道家对虚实空间的概念，在在都说明了空间概念在建筑历史中的演化过程。因此，我们可以说，建筑经常随着空间概念的改变而改变。20世纪末，人类的空间概念又随着网络的突然兴起而产生巨变，所谓的网络空间与我们的真实空间并存；又因为电脑模拟技术的发展，使许多不准备盖起来的建筑物，有机会在电脑上建造起来，成为所谓的虚拟空间。这些新的数码空间概念不仅影响设计的发展，也会全面颠覆我们既有的空间理论。建筑的空间概念将有机会像仪式空间、文本空间和表情空间一样丰富。

i-map / 信息地图

DESIGNER: Danijela Pilic + Barbara Leyendecker

i_map Concept   The idea of a museum goes back to the thought to conserve cultural and historical witnesses of bygone times for the future. The original carries all information and is cultural-historical record of the past. It's color, size and atmosphere can only be experienced with the original itself. When our world was more and more mechanized a demand for museums showing and explaining technical contexts emerged. Here the original is less important than the transport of information. Information becomes the exhibit.

i_map is a media based exhibition program that deals with the exhibition of information. Similar to 3D computer software, information maps are applied to walls in order to create a different space appearance. Virtual pictures become architectural elements in the real world of matter.

The focus is on the symbiosis of material and immaterial space: the exhibition, the software, cannot be experienced without a built museum, the appropriate hardware. Nevertheless i_map is not a classical architecture project but rather a concept that uses the interaction between virtual reality and our real world of matter in order to create architecture. i_map's contents, the software, the hardware and the resulting roomware are shown by means of a museum for astronautics.(At a museum for astronautics it is more important to be informed about the way a satellite works and about its impact on our lives rather than about the way it looks like. Accordingly an exhibition deals more with the software provided by a satellite than with a satellite itself, the hardware.)

i_map Software   Instead of determining an exhibition the i_map offers the possibility of every visitor choosing his own theme and navigating through the museum like through a three-dimensional web site.This kind of software based museum demands two different type of programs: a building system coordinates the user's input, calculates the individual ways and manages the submitted data. The exhibition software contains all projected exhibits, virtual realities and sound effects.

i_map Hardware   The exhibition works on three exhibition levels: every visitor gets information about his individually chosen theme at a personal level. Therefore every person gets a minicomputer at the entrance, called the personal guide. In small room units consisting out of holographic panes of glass visitors with similar themes are brought together to get common information. Projections in larger room units give information for a sizable group of people.

i_map Roomware   The technical character of the exhibition and its high degree of automation redefines classical museum areas. The lobby becomes the login where every visitor has to pass through several stations before he is led to the gateway, an area between login and exhibition that functions as a barrier between the outside world and the i_map. The exhibition is the interface that mediates between the concrete room and its possible outward form.

In putting virtual exhibits in a concrete room its outward form changes in our perception: a rectangular room occurs circular. The concrete room is different from the room a person perceives. Moreover virtual reality is not tate reality. By superimposing concrete space with virtual reality, new room structures may be created.

i-map 的概念   博物馆的最初设计理念是为未来保留过去的文化和历史的见证。保存下来的物件，记录着流逝的时光、带来所有的故事，只有通过对原物的体验，颜色、尺寸和氛围才得以显露。但当我们的世界越来越依赖机械，对科技脉络的展示和解释成了新需求。以往的原物、原件变得太不重要，原物的故事或信息、资料才是重点。

i-map 是一个以媒体为主，处理信息展览的展示计划；类似于电脑的 3D 软件，信息将像传统定义下的"墙"一样，用来塑造一个有别于以往的空间形态，虚拟的图像如真实世界中的建筑元素一样被运用。

这个计划的焦点在物质和非物质世界的共生：展览，即软件的部份，如果没有一个硬件（即实体）的博物馆来承载，也是无法被体验的。i-map 不是一个传统的建筑计划案，它更像是一个概念，利用虚拟实境和物质世界之间的互动，赋予建筑新的意义。而这次 i-map 所探讨的内容、软件、硬件、和展示空间，是通过一个以太空为主题的博物馆来呈现。（对这样的博物馆来说，比起展示人造卫星的实体，更重要是呈现它的功能及其对我们生活的影响。因此比起硬件来说，筹备一个关于人造卫星的展览，软件上面会存在更多的挑战。

COUNTRY: Germany / 德国

i-map 的软件　i-map 让每一位游客选择他自己的主题，而非明确地界定展览的范围。就像巡游在三维空间的网站中，引领自己遨游于博物馆。这类以软件为基础的博物馆需要两种不同的计划书：一是处理游客的进出、循环路线及资料管理的建筑物系统，另外则是包含所有主题展示资料、虚拟实境及声音效果的软件系统。

i-map 的硬体　展览的方式分三个层次：每位参观者可以依选择的主题而得到信息——这是个人的层次，每个人将在入口的地方拿到一台称为"个人游览器"的迷你电脑；在小一些的房间单元内，设有玻璃制的信息板，相近主题的参观者会被带入同一空间，体验共同的内容；而大展室中的资料则是提供给相当规模的人群。

i-map 的空间　这种展览方式的技术性和自动化程度，重新定义了博物馆在空间上的传统区分。大厅是每位参观者在进入入口前的数个登入站；介于登入区和展览区之间的入口则像是i-map和外面世界之间的一道栅栏；而展览则像是界面，在混凝土的空间和展览拟塑出来的环境之间充当媒介。

虚拟的展示环境会随着我们的认知而改变：一个矩形的空间可以变成圆的；参观者所体验到的空间和实际是不一样的。此外，虚拟实境并不只是用来重现真实世界，通过它的叠影效果，或许会产生新的空间架构。[翻译 洪士尧]

↓ VRML area: by superimposing concrete space with virtual reality new room structures may be created ／ 虚拟实境区域：通过将虚拟实境叠影到混凝土的空间上，也许会创造出新的空间结构

↑ Gateway: abstract impressions of space are supposed to get the user into the right mood for the exhibition ／ 入口：抽象的空间将引领参观着进入特定的氛围
↓ Every visitor chooses a personal topic and navigates the museum like a 3 dimensional web site ／ 每一个参观者选择个人的主题，像在三维空间的网站上，遨游这样的博物馆

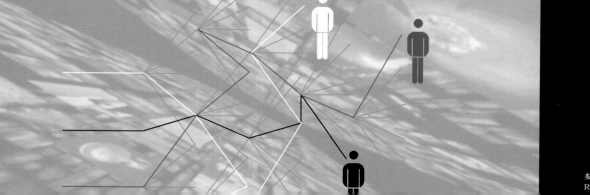

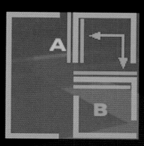

参与共同项目的使用者可以看到对方，但看不到对方的主题 ／ BELOW
RIGHT:Users joining a common event see one another but they

↑ Bitmap area: different exhibits require sophisticated forms of surroundings, i.e. the effect of a lift off increases in a very high room ／ 点状地图区：不同的展览需配不同的环境，例如在较高的房间即有提升的效果

↑ Haptical area: a climabox with different temperatures and otheratmospheric humidity ／ 合成照片区域：可转换不同温度和大气湿度的气候箱

Plötzlich taucht hinter dem Rande des Mondes in langen, zeitlupenartigen Momenten von grenzenloser Majestät ein funkelndes blauweißes Juwel auf: eine helle, zarte, himmelblaue Kugel, umkränzt von langsam wirbelnden weißen Schleiern. Allmählich steigt sie wie eine kleine Perle aus einem tiefen Meer empor, unergründlich und geheimnisvoll.

Suddenly from behind the rim of the moon, in long, slow-motion moments of immense majesty, there emerges a sparkling blue and white jewel, a light, delicate sky-blue sphere laced with slowly swirling veils of white, rising gradually like a small pearl in a thick sea of black mystery.

It takes more than a moment to fully realize this is Earth…home.

Interface / 界 面
DESIGNER: Aaron Cohen

The New Object / Interface Interface contains ways to communicate with both of them and it can be visualised as a filter- membrane, with two orientated sides. We have seen screen based interfaces and more complicated systems which work as described but the question is whether we can create architectural space? The aim of the project is to explore the new potential that new media bring in the field of space .

Our second interest in this project was the exploration of cyber_space which means land of control .Cyberspace is not only as they are distributed throught the internet, or installations in the enclosed space but also in the large scale content. Technological interventions can be applied at the natural land. This can not only expand the uses of technology into architectural reality- instead of the web expanding theoretical aspects of architecture- but also can be minimal, expressing the respect of our society to the natural environment together with the technological innovation.

Interface is a Field of Interaction and /or The Way of Interaction between 2 Poles.

The City & The Site The Site of the project is the Airport of Hellinicon located in the city of Athens, Greece. Less than 2 years Athens will have a new airport. The current airport will be abandoned, openning up an amazing space for architecture. There the city of Athens plans to create a kind of Metropolitan Park as a new use for the Airport space. As a place the site is located concessive topography as it is between the mountain and the sea.

The architectural structure of airports combines by default two different scales, the scale of the human and the scale of the aircraft. The site of the Athens airport can be read as real and virual at the same time: real because it exist, virtual because it is not easy to access it, real because it can be spoted on the map. Virtual because shooting pictures is prohibited. real because it is working today. Virtual because it is known that this use will stop, after the new airport construction will be completed , aproxximately in 2002.

Design Concepts Following the basic concept of the hangar we propose a re-usage of its shell for a public building. The aim of the building is to provide a center for experimentation, resourses and research if the field of art and technology for the city of Athens. Also to articulate an entrance into the Metropolitan Park in the former airport in a sophisticated way, re-using an existing airport building.

The new space of the hangar is still divided into two parts, two uses of different scale. We transform the older use of the labs into the real uses such as the media library , cybercafe and media laboratory, and the older use of the aircraft space into a virtual area. An area of experimentation, where the virtual meets the real,where the virtual can exist because of the real support.

新物件和界面 界面是两极间互动的区域或沟通的方式，就如同一个两面的过滤膜。这种功能我们在界面网以及更复杂的系统中曾经看到过，我们看过这些系统，但接下来的问题是我们可以创造建筑空间吗？本方案的目的便是开拓新媒体对于空间领域的新潜能。

本方案的第二个目的在于拓展虚拟空间，意为控制地盘。虚拟空间并不只分散于网络或有限的空间内，也分布于大范围的空间中。自然土地上可以使用技术性上的创新，除了代替网络拓展的建筑理论将技术运用在实际建筑上之外，至少可以表示社会对自然环境与技术发明的重视。

城市与场地 本方案的场地，Hellinicon机场位于希腊雅典。雅典将在两年内拥有新的机场，目前的机场将会被废弃，为建筑开启惊奇的空间。雅典城打算建立类似市民公园的机场空间，场地处在山与海中间的地带。而机场的建筑结构自动结合两种不同的尺度，即人类与飞机的尺度。

雅典机场的场地可说同时具有真实与虚拟两种性质：因为存在而真实，因为不容易接近使其虚拟；因为可以在地图上点出而真实，禁止拍照使其虚拟；因为正在使用而真实，大约在2002年，新机场完成后将会停止使用旧机场的从而使其虚拟。

COUNTRY: U.S.A. / 美国

设计概念 类似于飞机棚的基本概念，提出将壳重新使用在公共建筑上的想法。建筑物的目的在于提供一个雅典城艺术与技术测试、资源与研究中心，入口自然的面向旧机场都会公园，重新利用旧机场建筑。

飞机棚的新空间依旧分成不同尺度的两部分，原来的实验室转变为媒体收藏馆、网络咖啡以及媒体实验室，而原来的飞机区则成为虚拟区域；至此形成虚拟与实际交错的区域，一个由实际支撑、同时又存在着虚拟的区域。因而此建筑可解释为一套存在于飞机棚内的物件。[翻译 王俐文]

↓ 室内电脑3D模拟图 ／ The 3D computer-simulated diagram

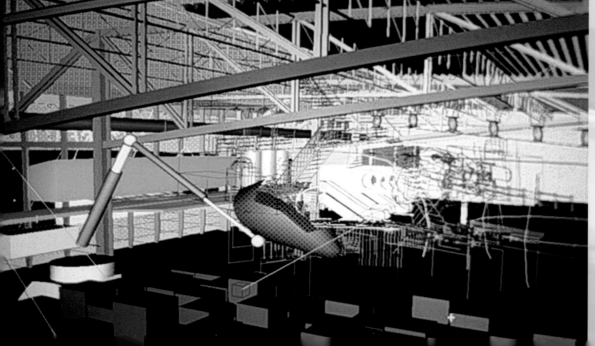

↑The 3D computer-simulated diagram ／ 室内电脑3D模拟图　↓Perspectives ／ 透视图

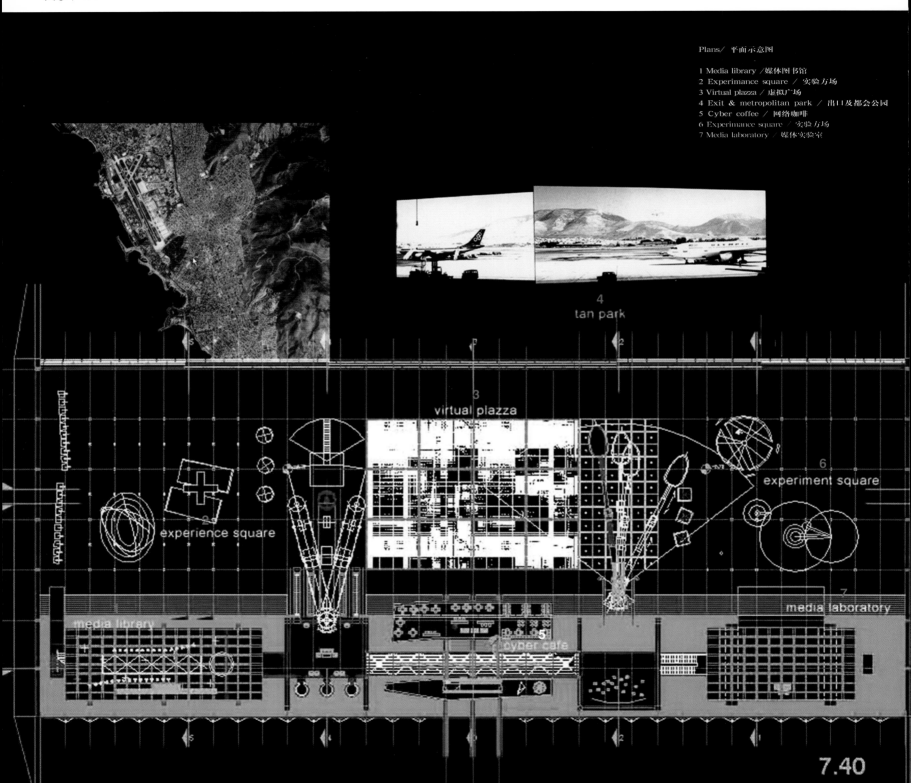

virtual plazza

experience square

iNSTANT eGO / 瞬间的自我

Adrien Raoul + Hyoungjin Cho + Remi Feghali

Concept   In *"La Ville, Territoire des Cyborgs"*, Antoine Picon introduces cyborgs(a cyborg is a mam/machine.) who roam in the urban territory with a constantly growing speed. Interruption time, non-time is thus multiplied and made inevitable. Our nomadic space, produced by these intersticial durations, allows the emergence of intimate electronic space in the middle of the urban public space. Plugged to our clothing, iNSTANT eGO is primarly a cluster of intelligent tissue folded over, waiting to be unfolded. When released, the interface unfolds, subjective time is trapped in our personal space, creating an ambiguous space-time relationship where other people's time is only subject to our personal intimate time. Space is transformed by its own kinematics, headed towards the time "zero", where telecommunication is the main source of pictural and lingual information: surfing the net, thinking, writing, refreshing ourselves, we are capable of achieving our most intimate gestures in the middle of the urban realm. This plugable and flexible space interacts indirectly with its close environment. Parametrable, iNSTANT eGO is sensually related to the local context, transmitting its energy flows, its virtual potentiality.

概念   在《城市，电子人的领域》一书中，Antoine Picon 所描述的电子人（注：或称半机械人）以持续增长的速度漫游在城市之中。被切割的时间、漂浮的时光越来越多；而我们流浪的空间，从缝隙里生出，使一种私密的电子空间得以浮现在城市的公共场合中。"瞬间的自我"原本是一串折叠起来的智慧薄膜，附挂在我们的衣着上，等待着被释放。当开启时，膜面鼓张，主观的时间感被包覆在个人的空间中，一个暧昧的时空关系产生。他人的作息只是我们私密时光的从属。而空间被自身的动力逻辑转化，导向"零时"：电信通讯是图像和语言交流的一种主要方式，如上网、输入、思考、反省等。即便在城市的最中心，我们也能完成自身最亲密的举止。这个可连接而且富有弹性的瞬间自我，并不直接和周边的环境互动，而以感性的方式和当地的文脉发生关系，输出能量和它的虚拟潜力。[翻译 洪士尧]

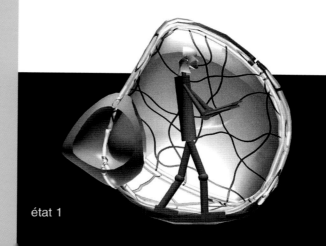

état 1

Country: France / 法国

↑ The plan from folded to unfolded收放平面图

↑ The section from folded to unfolded收放剖面图

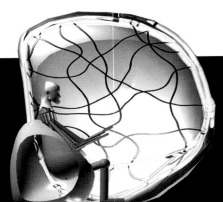

↑ Textile screen ／ 织状幕面   ↓ The computer–simulated diagrams of folding process／ 收束状态电脑模拟图

état 10

état 0

ETAT 0 (DEPLACEMENT)

Accrochage au vetement
par fermeture eclaire

iNSTANT eGO REPLIE

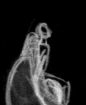

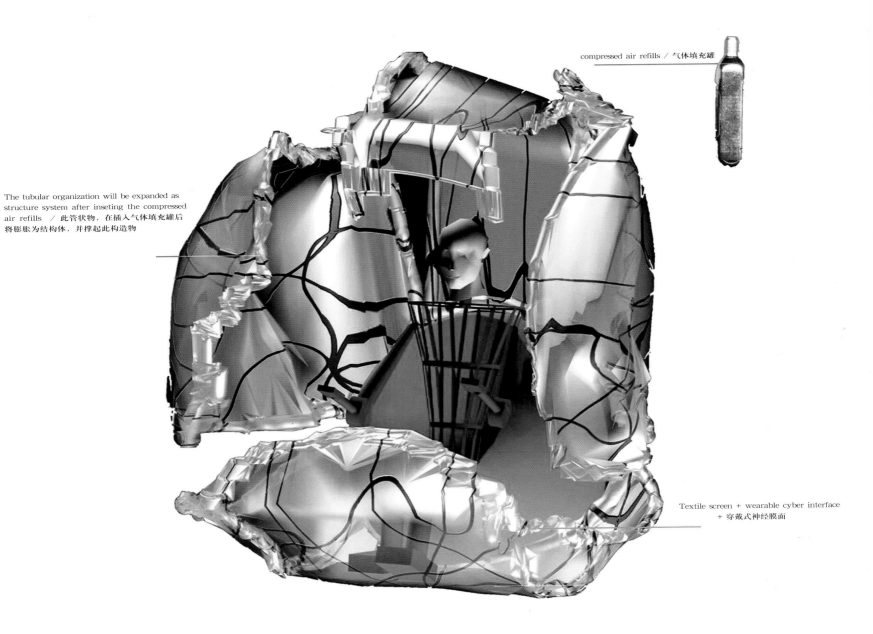

compressed air refills / 气体填充罐

The tubular organization will be expanded as structure system after inseting the compressed air refills / 此管状物，在插入气体填充罐后将膨胀为结构体，并撑起此构造物

Textile screen + wearable cyber interface + 穿戴式神经膜面

↓ The conceptual sketches / 概念发展草图

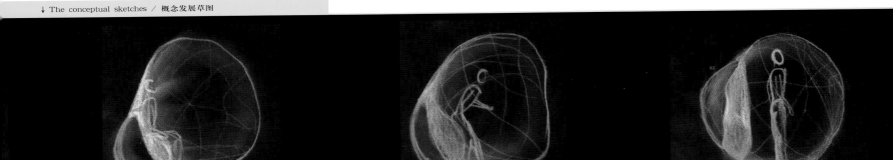

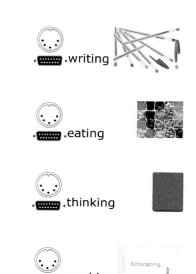

.writing

.eating

.thinking

.washing

The functions of iNSTANT eGO: writing, eating, thinking, washing, resting, dressing, connecting, discussing / 瞬间的自我之功能：书写、饮食、思考、洗涤、休憩、衣着、连络、讨论↑

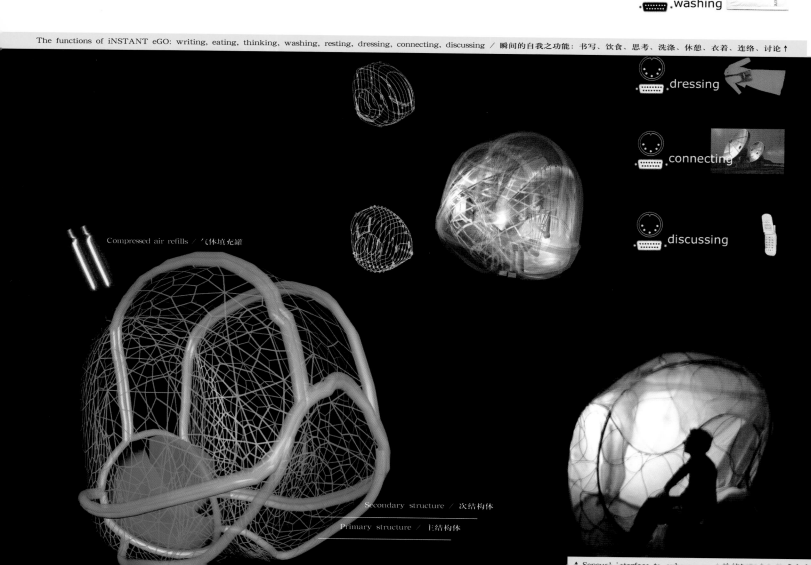

.dressing

.connecting

.discussing

Compressed air refills ／气体填充罐

Secondary structure ／ 次结构体

Primary structure ／ 主结构体

Beyond the Year 2050 - Definitions of Micro- and Macrospace Based on the Project of Manned Research Space Craft
微空间与巨空间的定义

DESIGNER: Agnes Zwara

Is Digital Architecture All In One or One In All?—[Future belongs to bio-tech cells]

For ages mankind has been perceiving their environment, in terms of local relations, in the limited extent. Restrictions that did not allow them to perceive it in the wider context were obviously observational limits and scientific ones.That perception, in a way, determined aesthetic and functional values of the objects which were created by people (from adapting natural goods, building medieval castles that were not intruding but coexisting with natural environment, and finally to deep interfering in nature in the industrial era). Scientific progress has enforced the necessity of changes in interpretation of mutual relations between artificial objects made by homo - sapiens and nature.

Discovery made by Copernicus proved that the Earth was not the center of the Universe. It initiated revolutionary changes in notional interpretation of various branches of human existence. However, it did not change the architectural approach. Architecture remained submitted to ancient demagogy - ancient idea controlling the Universe. Therefore, architecture embodied misunderstood Aristotelian "figure one" idea. The idea itself, however, referred only to the Earth. It did not initiate the necessary dialogue with the Universe.

Through ages architects interfered in nature taking into consideration only its one dimension - the space context. They did not perceive nature as a whole - the complex of time, substance and its acting. Therefore, conceptualistic and philosophical bases remained far behind all the branches of knowledge that were developing intensely during that time. They, although not technological, were strongly connected with those bases.

Fragmentary observations has led to false conclusions dependant only on the subjective point of view. In micro - scale one may observe pure functionalism without the space divisions. Functionalism changes its character depending on the surrounding and, at the same time, it influences the surrounding. In macro - scale one may observe the unity in relation to the Universe. However, the object preserves its own micro - scale. The consequence of global micro - and macro - mutual relationship approach is not the result of basic algebra equation. Such an attitude diminishes the gap that has appeared suddenly in the of permanent digital evolution process. The lack of updated definition of the term "architecture" and its activities leads to further penetration of that area.

Mankind has integrated their way of thinking and acting with silicone machines. Since man started "thinking with their fingers" the new historical era of interactive - digital culture has opened. Evolution in the interactive media and digital technology cast the new light onto the human psychology and other relations between Man and Space. Real and virtual spaces intermingle each other. They create new reality, new definition of presence, consciousness and acting.

It is quite likely, that we may soon be faced with the results of an open competition that takes place between purely technocratic and techno-gnostic current of changes; the results of which may possibly have an influence on defining what digital architecture really is. For the time being, it is hard to say whether the term "digital architecture" applies more to something existing solely in the form of a computer code, than to what has been generated by digital means, or perhaps to something equipped with technologically advanced media. At this stage of technological development, it is feasible to imagine a future, where a human living withing a virtual world will, depending on his personal preferences, encode its characteristics.

Up till now it was possible to show discrepancies in emotional reception of the same space by various humans but in future general perception of space will cease to be a common matter for a specific population - it will be each observer's matter in all of its aspects. The further natural evolution of structures, social dependancies and even human mentality seem unstoppable in their direction set by technological current. The new approach with the new order makes us seek the matrixes, alternative solutions that would allow the use of all the media in the processes of design and usage. Referring to the above - mentioned theory of the micro- and macro-universes' correlation one should draw the conclusion that the optimum solution in design is to use functional combination of nature and technology. The latest technological discoveries allow a full-scale exploration of space. Recent manned expeditions are a good example thereof. The greatest obstacle in space research so far, have been the financial requirements, especially costs of setting up space stations on the orbit.

COUNTRY: Poland/ 波兰

数码建筑代表所有的建筑，还是建筑中的一种？（未来在于生物科技）

人类一向顺应环境，不论是地方文化还是外在条件。这些限制都阻碍了人们对环境进行更深的了解，无论在观察上还是科学上。无论在美学上还是机能上，这种认知也限制了人们的创造（如中世纪的城堡，它其实与自然共存，并无法侵入，违背自然；而到了工业时代，恰是干预自然）。科学进步，反而加强了人造物间的关系—空间与自然。

哥白尼的发现证明了地球并非宇宙的中心，从此改变了人类观念上的认知，对建筑却无多大的影响。因为建筑善于蛊惑人心—自古即为控制宇宙的媒介。建筑实际上成了亚里斯多德常遭人误解的"形体"理论的表现，此理论其实指土地，而非与宇宙的对话。

长久以来，建筑师总以空间内容来考虑自然，而忽略了自然的整体性—包括时间的存在、作用与它的复杂性，不像其他学科，建筑的观念与哲学虽然无关，却也未随时间快速发展。

关于建筑的片断式观察，常导致较主观的结论。如以较微观尺度来观察，会得出一个机能主义的观点，而无空间的分隔，机能主义会随环境变迁，同时会影响环境。就宏观的尺度而言，应观察与宇宙关系的一致性，但被观察者仍会保持其本身的微观尺度。这种全球性的，微观与宏观间的相互关系，可以用单纯的公式来解释，此种态度与哲学，亦可消弭数码化中的片断。如果建筑的定义并未随时代更新，势必会导致更大的落差。

The latest results of astrophysical researches seem definately optimistic, as they allow for plannig a research station, to be built on the Moon's surface (or its orbit). The existence of such facility would make provision for executing cheaper space expeditions (due to weaker gravity and smaller force necessary to leave its gravitational field) and could be a site coordinating transfers in space travel or technical HQ, specialised in assemby of manned space stations brought from Earth in form of kits. The enterprise itself seems unreal enough for laymen, but there are serious plans for fulfillment of these. Cost-cutting implies building space stations permanently suspended upon orbits, which in turn determine technical specifications of a satellite. What is more, such specifications will be considerably influenced by the outside conditions.

Main functional systems presented in the research space craft are technological transposition of the adequate vital functions of Daphnia: air conditioning = respiration system, double water supply system = blood - vascular system and energy supply (solar batteries) = photosynthesis. Holistic perception of the object is possible in The Universe which allows for many ways of creating visual reception of the form.

The following design comprises 3 possible textures of research station's solid: [1]TECHNIC — the selection of colours consists of natural qualities of materials used in creation of facilities of such kind. A steel colour is dominating. Such selection makes a station have a neutral influence on the surroundings. [2] HIDDEN —a camouflage version, taking advantage of reflective qualities of materials used to shield the fuselage. The fuselage reflects surrounding elements, hence becoming an integral part of the environment. Being naturally merged with, it becomes indiscernible for an observer. [3] ATAVISTIC — radically different to the aforementioned, based upon a contrasting range of colours systematized with respect to functional division of separate elements. Such selection of colours will also resemble atavistic warning hues, used by amphibians to repel their natural enemies.

The perception of the station will differ, depending on the observer's position (images 23(25). At a large distance, the beholder will have an impression that he is observing a microorganism through a microscope. For an inhabitant it shall feel as an entire environment, where he can exist undisturbed.

The ideas borrowed from nature are not literal. In fact, they show only the activity patterns based on contemporary technological progress. The basis for such ponderings is their relativity, which alongside "plasticity" of thought-creation can be regarded as a fundament of my understanding of "digital architecture".

However, no matter which option we take into consideration, the ship craft being individual and independent object confirms false dualistic, already existing, nomenclature in architecture.

由于电脑的发展，这些硅晶片已逐步影响人们的思想，现在人们用手指敲打键盘来思考，于是新的数码文化时代开始了。互动式的媒体与数码科技，开启了新哲学及人与空间的新关系。真实与虚拟互相混合，创造了新的真实、新的定义、知觉及行动。对于新秩序的追求，会带来新的矩阵、新的解决方案，并会在设计及使用上采用各式媒体。

综观以上的论述，包括微观、宏观宇宙间的关系，我们可以总结，就建筑而言，较乐观的解决方案仍旧是机能与科技的结合。在这些研究空间中，重点是科技的过渡：空调类比排汗系统；双层墙类比血液与血管系统；能源系统则运用光合成的原理，从自然中，我们仍能引入许多原理。我们也需要以当代技术来建立动态模式。

无论如何，不管我们采取何种方向，这些技术都会独立于建筑以往的双重性、专业性及独立性。[翻译 赵梦琳]

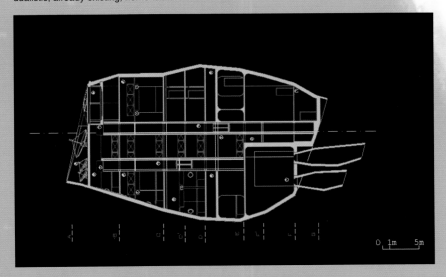

剖面图／ Section

a　Telescope Level
b　Astrophysics Labs Level
c　Bio-geo-ciem Labs Level
d'　Living Area
d　Generalship Level
e　Isotopic Labs Level
f'　Steer Level
f　Engine and Water Technology Level
g　Exit Level

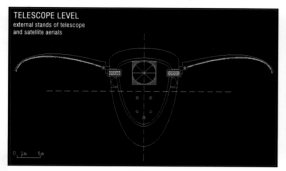

TELESCOPE LEVEL
external stands of telescope
and satellite aerials

0　1m　5m

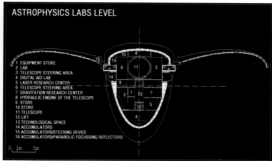

ASTROPHYSICS LABS LEVEL

1 EQUIPMENT STORE
2 LAB
3 TELESCOPE STEERING AREA
4 DIGITAL AID LAB
5 LASER RESEARCH CENTER
6 TELESCOPE STEERING AREA
7 GRAVITATION RESEARCH CENTER
8 HYDRAULIC ENGINE OF THE TELESCOPE
9 STORE
10 STORE
11 TELESCOPE
12 LIFT
13 TECHNOLOGICAL SPACE
14 ACCUMULATORS
15 ACCUMULATORS/STEERING DEVICE
16 ACCUMULATORS/PARABOLIC FOCUSSING REFLECTORS

0　1m　5m

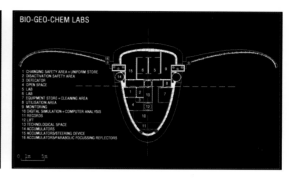

BIO-GEO-CHEM LABS

1 CHANGING SAFETY AREA + UNIFORM STORE
2 DISACTIVATION SAFETY AREA
3 DEFECATOR
4 OPEN SPACE
5 LAB
6 LAB
7 EQUIPMENT STORE + CLEANING AREA
8 UTILISATION AREA
9 MONITORING
10 DIGITAL SIMULATION + COMPUTER ANALYSIS
11 RECORDS
12 LIFT
13 TECHNOLOGICAL SPACE
14 ACCUMULATORS
15 ACCUMULATORS/STEERING DEVICE
16 ACCUMULATORS/PARABOLIC FOCUSSING REFLECTORS

0　1m　5m

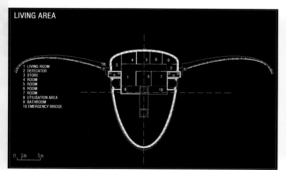

LIVING AREA

1 LIVING-ROOM
2 DEFECATOR
3 STORE
4 ROOM
5 ROOM
6 ROOM
7 ROOM
8 UTILISATION AREA
9 BATHROOM
10 EMERGENCY BRIDGE

0　1m　5m

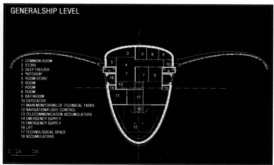

GENERALSHIP LEVEL

1 COMMON-ROOM
2 STORE
3 DEEP FREEZER
4 "KITCHEN"
5 ROOM STORE
6 ROOM
7 ROOM
8 ROOM
9 BATHROOM
10 DEFECATOR
11 MAIN MONITORING OF TECHNICAL TASKS
12 NAVIGATION/FLIGHT CONTROL
13 TELECOMMUNCATION ACCUMULATORS
14 EMERGENCY SUPPLY
15 EMERGENCY SUPPLY
16 LIFT
17 TECHNOLOGICAL SPACE
18 ACCUMULATORS

0　1m　5m

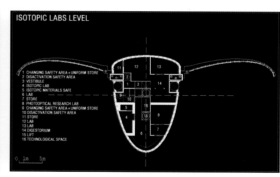

ISOTOPIC LABS LEVEL

1 CHANGING SAFETY AREA + UNIFORM STORE
2 DISACTIVATION SAFETY AREA
3 VESTIBULE
4 ISOTOPIC LAB
5 ISOTOPIC MATERIALS SAFE
6 LAB
7 STORE
8 PHOTOOPTICAL RESEARCH LAB
9 CHANGING SAFETY AREA + UNIFORM STORE
10 DISACTIVATION SAFETY AREA
11 STORE
12 LAB
13 LAB
14 DIGESTORIUM
15 LIFT
16 TECHNOLOGICAL SPACE

0　1m　5m

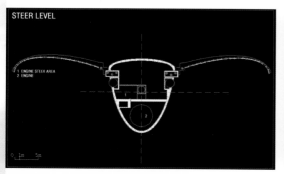

STEER LEVEL

1 ENGINE STEER AREA
2 ENGINE

0　1m　5m

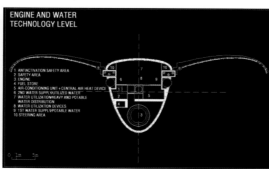

**ENGINE AND WATER
TECHNOLOGY LEVEL**

1 ANTIACTIVATION SAFETY AREA
2 SAFETY AREA
3 ENGINE
4 FUEL STORE
5 AIR-CONDITIONING UNIT + CENTRAL AIR HEAT DEVICE
6 2ND WATER SUPPLY/UTILIZED WATER
7 WATER UTILIZATION/HEAVY AND POTABLE
   WATER DISTRIBUTION
8 WATER UTILIZATION DEVICES
9 1ST WATER SUPPLY/POTABLE WATER
10 STEERING AREA

0　1m　5m

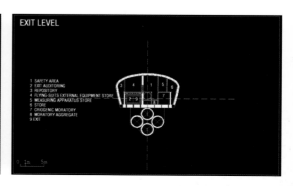

EXIT LEVEL

1 SAFETY AREA
2 EXIT AUDITORING
3 REPOSITORY
4 FLYING-SUITS EXTERNAL EQUIPMENT STORE
5 MEASURING APPARATUS STORE
6 STORE
7 CRIOGENIC MORATORY
8 MORATORY AGGREGATE
9 EXIT

0　1m　5m

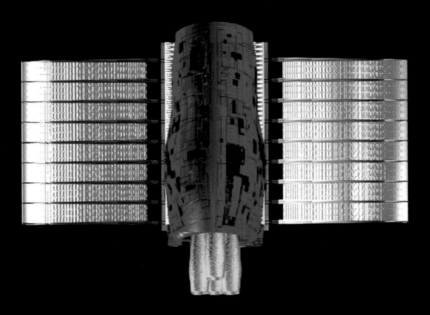

↑隐藏—第二个变化的特色，是反映太空船表面的样子，能够在无际的宇宙中隐藏外表。在环境中观测站在视觉上主要的冲击是无形的。
Hidden — 2nd variant-based on the characteristic reflective features of the spacecraft's surfaces which make possible to be visually hidden in the dark open sphere. Main visual impact of the station on the environment is invisible.

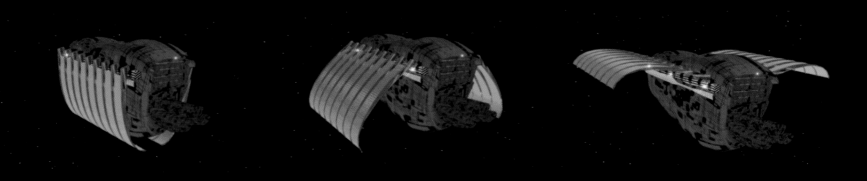

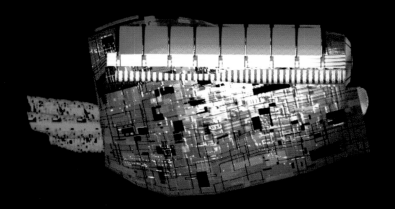

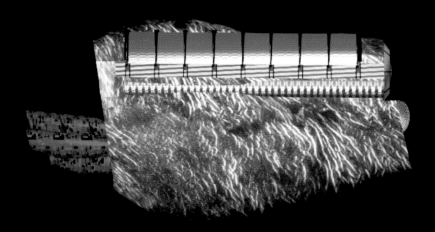

↑科技——第一个变化是以建造太空船材料的原色，主要是铁灰色。在环境中观测站在视觉上主要的冲击被认作为无确定性质的。/ Technical 1st variant-based on natural colours of material used to build the spacecraft/ mainly gray-steel colour palette/. Main visual impact of the station on the environment is considered to be neutral.

↑重回生物特性——第三个变化的特色，类似一些两栖动物生物学的特征。亮丽的色彩是对危险的警讯。在环境中观测站在视觉上主要的冲击是在威慑侵入者/ Atavistic — 3rd variant-based on the characteristic atavistic biological features of some groups of amphibian. The bright colors palette warns of a danger. The main visual impact of the spacecraft on the environment is to repel and to frighten intruders.

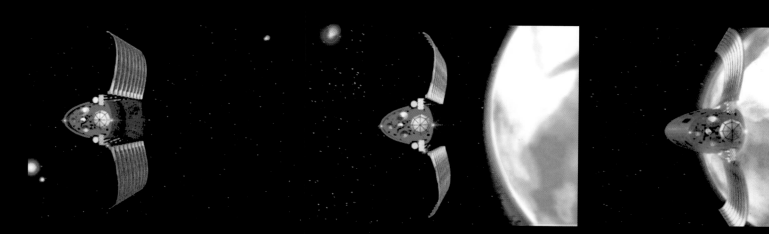

Stratapolis — Third Millenium Cities, 4 Hypothesis / 第三千禧年城市

DESIGNER: Castelli Didier

With the beginning of the third millenium, the City which didn't completely sustain the mutations of the society evolutions, can't develop based upon the models of the second millenium. The City has to morph into an other entity.

4 hypothesis about 4 thems of development are suggested.

[1] **Urban Expansion** With more than 6 billion humans beeings, the City can no more expand lawlessly. Regulated by a " theoretical limited skyline ", the future City will develop a stratum system in altitude.

[2] **Urban Structure** With energy and pollution problems, the City will develop new structures. Each block becomes his own autonomous factory which manages the cycle of production, consumption and wastes.

[3] **Urban Communications** With the computer , the City which can no more assimilate the physical means of transportation, develops other virtual communications for all activities and finally liberates the urban space.

[4] **Urban Spaces** With the end of street and corner concepts, the City develops new public spaces in the old urban empty spaces and in the new urban spaces producted between the new strata.

当第三个千禧年到来时，第二个千禧年建立的城市已不能满足其需求，城市势必在本质上有所渐变。

以下为本设计的四个假设:

[1]城市扩张 人口超过六十亿时，城市便不能毫无限制地发展，我们将以天际线的效果做为其扩张的依据。

[2]城市结构 城市有新的空间、结构来解决能源及污染的问题，每个街区都是自给自足的单位，自行解决垃圾，自行生产所需的物品。

[3]沟通 随着电脑的普及，城市不再依赖运输系统，而会替代以其他的沟通方式，如此可解放城市空间。

[4]城市空间 街道与街角将会消失，在老城区的空地和新城区新阶层的空间中将开发出新的公共空间。[翻译 赵梦琳]

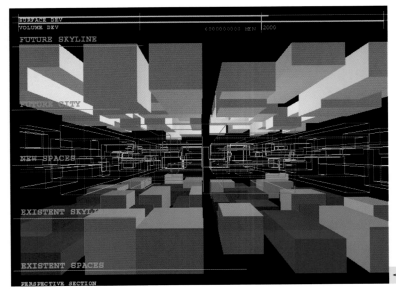

← Urban expansion, conceptual scheme/ 城市扩张

AFFILIATION: Architect epfl
COUNTRY: Switzerland/ 瑞士

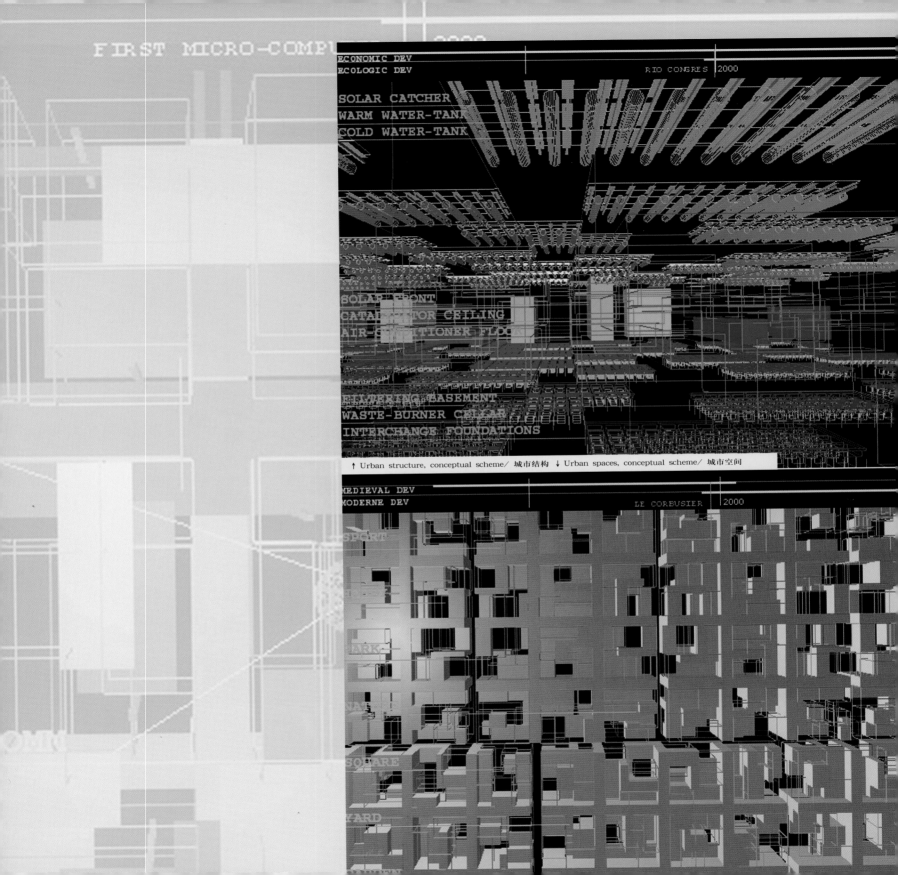

ECONOMIC DEV
ECOLOGIC DEV
                                                                RIO CONGRES   2000

SOLAR CATCHER
WARM WATER-TANK
COLD WATER-TANK

SOLAR FRONT
CATALYSATOR CEILING
AIR-CONDITIONER FLOOR

FILTERING BASEMENT
WASTE-BURNER CELLAR
INTERCHANGE FOUNDATIONS

↑ Urban structure, conceptual scheme／城市结构 ↓ Urban spaces, conceptual scheme／城市空间

MEDIEVAL DEV
MODERNE DEV
                                                                LE CORBUSIER   2000

SPORT

MARK

SQUARE

YARD

↑ Communication spaces, theoretical view／空间

Architectural Design of a Virtual Campus/ 虚拟校园的建筑设计

DESIGNER: Ning Gu + Mary Lou Maher

The design of a virtual campus, while considering the role of technology in education, also can be influenced by the sense of place that more traditional architectural design provides. This approach need not result in a virtual campus that looks like a physical building, the main concept is to intricately link learning with an experience in a place with other people.

The design project described here is in response to the need for a third and virtual campus for ETH in Zurich. Our design, while using place as the basis for the learning experience, organises the various places into three distinct functional spaces: the administrative space, the academic space, and the professional technology transfer space. These three spaces form the organisational backbone, bringing together the physical campus, the virtual campus and the technical infrastructure supported the campuses.

The design concept includes a 3D Virtual World as a place for building the community through meeting people and as a 3D space for accessing the variety of information repositories. Often, the use of internet technologies results in "downloading" information and managing information navigation. In other words, the person is not part of the information and the information is sent out to each person's personal computer. Our design concept is to bring the community in to the campus to access information and people. The technical infrastructure supports this concept by including a representation of the individuals that are part of the community and organising information in a variety of technology formats to facilitate information handling, creation, maintenance, and access.

The plan view of the Virtual World is circular, with curved glass panels creating three ovals that are indicative of the boundaries of each space. Providing glass panels for the boundaries allows a person to see other parts of the virtual campus regardless of their location. The boundaries of the separate spaces provide levels of privacy, security, and interactivity through their implementation as software agents. We intend that the objects in the virtual world carry a metaphorical reference to the physical world while being made of software modules.

In the design concept, the use of shape is consistent and intentional: curved shapes indicate movement and rectangular shapes indicate stability. For example, portals and corridors are curved and an office is rectangular. We use open frame walls to provide a visual boundary that can also be functional as a frame for holding information and tools. Sound design is also a consideration. Sound in a physical world happens naturally. In the virtual world, the sound is designed to create ambience and to provide information.

Within the Academic space, there is a structure modelled to be similar in style to the buildings in the Honggerberg campus at ETH. The buildings that house the Faculties, Departments and offices are rectangular and stable. The portals and the corridors have curved shapes indicating movement. Access to the physical campus through video windows provides access to activities occurring in the physical campus through the virtual world. Within the Administrative space, the curving walkway with circular spaces provides access to enrolment and administrative information. Management staff also have access to manage the campuses virtually. A stylised representation of the Visdome at ETH provides a 3D space for virtual presentations and demonstrations. The Professional space provides a virtual

虚拟校园的设计在考虑科技在教育中扮演的角色时,也会被传统的建筑设计方法所提供的意识影响。我们不必创造出向实体建筑看齐的虚拟校园,而应是连接学习以及和他人相处的场所。

所做的方案适合了苏黎世对第三个虚拟学校的需求。设计在将场所视为学习经验的基础的同时,将不同的场所组织为三个特定的机能空间:行政空间、学习空间与专业技术转移空间。这三个空间形成编制的骨干,结合实体校园、虚拟校园与技术性公共设施来支持这个校园。

设计理念包括作为组织社群的3D虚拟场所以及使用多样信息资料库的3D场所。虽然信息传送到每个人的个人电脑中,但这个人自身并非信息的一部分。设计概念是将社群带入校园以获得信息与人群的接触。通过将社群个体的代表与技术编排组织资讯,使之易于操作、创作、维护和使用而技术的公共建设也支持了此概念。

虚拟世界的配置是圆形的,加上弧形玻璃面板构成三个象征各个空间界线的椭圆。玻璃面板的界线让人不管在校园何处都可看见虚拟校园的其他部分。个别空间的界线以软体中介的角色提供不同程度的私密性、安全感与互动。虽是由软件模块构成,但我们仍然打算让虚拟世界中的物件隐喻性地包含对现实世界的参照。

设计概念中对于形的使用始终如一而且具

AFFILIATION: Key Centre of Design Computing and Cognition, Faculty of Architecture, University of Sydney, NSW, Australia
COUNTRY: Slovenia/ 斯洛文尼亚

display area for professionals to participate in the activities of ETH. This is equivalent to a physical conference facility or a suite of professional offices in a physical campus.

The technical infrastructure provides data to support all campus activities, both for the physical campus and the virtual campus. The infrastructure is divided into three categories of information representation: the World Wide Web for integrated multimedia information, documents for reading material intended to be printed, and databases for structured information storage and retrieval. Each of these different types of information repositories have their own portals for access and modification. Casual access is through the Virtual World. Authorised access to make changes to the information is provided through special portals into each repository.

The implementation, similar in many ways to the construction of a physical campus, is an object-oriented persistent database of the world that can adapt to any new technologies in supporting virtual communities, communication, and interaction. The object-oriented database represents the properties and methods of an immersive 3D world with people represented as avatars; similar in concept to Active Worlds (http://www.activeworlds.com) or Microsoft Vworlds (http://www.vworlds.org). Each object in the database is an object in the Virtual World, with a 3D model for visualisation and properties and methods to represent the object and its interactivity with other objects in the world. The basic classes of objects include: rooms, portals, things, and people. These classes are extended and specialised to provide an implementation of the virtual world where each object has intentional functions, behaviours, and structure. The virtual world object database is linked to the World Wide Web, the documents, and the other databases.

In summary, the virtual campus design not only combines the physical and the virtual through the 3D space provided for the community access, it provides a particular style and presence for a sense of place that can change over time to reflect changes in the activities and people.

有涵义：圆弧形代表运动而矩形代表稳定。例如，正入口和回廊是弧形的，而办公室是矩形的。我们使用开放的框架墙提供视觉界线，同时也做为放置信息和工具的框架。我们也考虑到声音设计，在现实世界，声音自然地发出。在虚拟世界中声音设计出来是为了创造氛围与提供信息。

公共设施、各系和办公室的建筑都是矩形并稳定的。主入口和回廊圆弧代表动态。通过在虚拟世界传输现实世界所拍摄的影像，提供接触真实世界发生的活动。在行政空间中，圆形空间中的弧形走道形成了入学和行政管理信息的接触空间。管理工作人员同时可通过虚拟的方式管理校园。格式化的视觉穹窿代表虚拟展示与宣传的3D空间。专业空间提供专业人员参与ETH活动的虚拟展示空间。这和实际校园的实体会议设施或专业办公室是等同的。

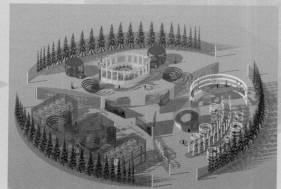

The virtual campus as three functional spaces in a virtual world. Each functional space is occupied by people as avatars and provides access to information and events./ 虚拟校园有三个机能空间。每个空间被具体的人使用，提供信息与活动。

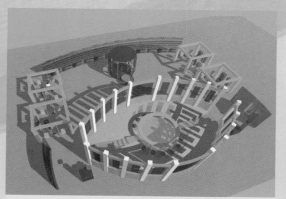

A virtual classroom that provides a focal point for discussion and the presentation of slides or shared drawing boards./ 一个提供讨论用焦点与幻灯片投放或共用绘图板的虚拟空间。

The virtual office area has unlimited sp/ 虚拟办公室的区域有无限的空间好让社区的每个成员拥有私人的虚拟办公室。

公共设施提供支持整个校园活动的信息，同时也为实体与虚拟校园服务。它被分成三个信息目录：提供给一体的多媒体信息的全球信息网络，可列印出来的阅读教材，提供给信息储存与读取的资料库。它们拥有各自供进入与编辑的主入口。

完成的结果与实际校园的建造有许多相似之处，是可以适应任何支持虚拟社区、沟通与互动新科技，并且是以物件为目的持续发展的世界资料库。它代表了一个沉浸的 3D 世界；类似Active Worlds(http://www.activeworlds.com)或 Microsoft Vworlds (http://www.vworlds.org)的概念。资料库中的每个物体都是虚拟世界中的一个物件，有模型呈现其视觉效果，有代表它以及它与世上其他物体互动的特性与方法。物件基本的分类有：房间、入口、物品与人。这些分类延展并专门提供虚拟世界的建筑结构，存在于其中的每个物件都有特定机能、行为和结构。它与全球信息网、文件或其他资料库连结。

虚拟校园的设计不只提供给社群使用的3D空间结合实体与虚拟，它亦提供了反映人在活动变化时，可随时间改变场所感的特定风格与外观。[翻译　蔡咏岚]

↑ The main entrance to the virtual campus places the person in the center with views to all three functional places. The glass partitions indicate the transition from one place to another and the information kiosk at the entrance provides instructions for the use of the virtual campus.／ 虚拟校园主入口将人置于可见到三个机能空间的中心带。玻璃面板象征从一处到另一处的过渡，入口处的信息亭提供了虚拟校园的使用方法。

↓ A virtual meeting point set up in one of the departments as a public place for the campus community. The glass panel can also provide video conference facilities.／ 一个建于其中一个学系的虚拟集合场所是校园社群的公共空间。玻璃面板也提供视讯会议的设施。

Infotube/ 信息圆筒

DESIGNER: Fumio Matsumoto + Shohei Matsukawa + Akira Wakita

Migration of the Concept of Architecture — Industrial revolution had driven people and goods to cities and given birth to forms of new cities and architecture. Information revolution accelerates flow of people and information into the network, and is creating a new electronic social space (or cyberspace). Buildings and cites in the real space can not move into the information space. But the concept of them can. INFOTUBE is the reconstruction of the real town scape (shopping streets in Yokohama, Japan) as a cyberspace architecture in an entirely different form.

Visitors to INFOTUBE can cruise around a new information space reconfigured into the cylindrical tube. This Tube is covered with lots of rectangular "cells," a minimum unit of information, on which various images and texts are displayed. Visitors can interact with these cells, get detailed information, and upload their own messages to the cells. It is also proposed to place "Live Cells" on the city streets by which the real space can have close interactions with cyberspace.

Information is distributed not hierarchically as in the typical web pages, but simultaneously and randomly on the Tube. So visitors can overlook information at one time without browsing deeper layers of the site. By spatializing and visualizing information, visitors in cyberspace could recognize information more intuitively just as when they are viewing landscapes.

建筑概念的迁徙　　工业革命使人口与货物集中于都市,催生新形态的都市与建筑。信息革命加速了人口与讯息流入到网络中,造成新的电子化社会空间(或虚拟空间)。现实中的建筑与都市无法在信息空间中移动，但是建筑与都市的概念则可以。信息圆筒是全然不同的形态，再现真实城市景观(日本横滨的购物商街)的虚拟建筑空间。

信息圆筒的访客可以在一个重新装配进圆筒的信息空间中漫游。这个圆筒覆盖着许多展示多样影像与文字的长方形单元，它们是信息的最小单位。访客可以在此获得详细信息及留言，与这些信息单元互动。另外甚至有将"活单元"安放到都市街道上的提议，这样真实空间将与虚拟空间交互作用。

信息并非像网页一般有等级分布，而是随机任意地放置在圆筒上。这样一来访客不用浏览网页更深层级的部分就可获得信息。将信息空间化与视觉化后，虚拟空间的访客将可如同观赏景观一般，更直觉地识别信息。 [翻译 蔡咏岚]

←Interior/ INFOTUBE 全景　→Exterior/ 簇群的圆筒

**AFFILIATION: Plannet Architectures + 000 Studio
+ Graduate School of Media and Governance, Keio University
COUNTRY: Japan / 日本**

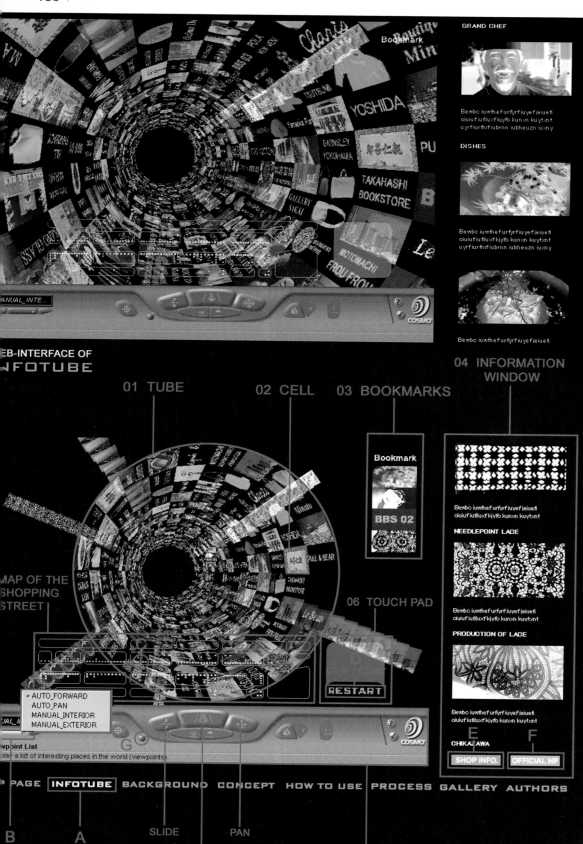

←Interface／ INFOTUBE 的使用者界面

←How_To_Use／ INFOTUBE操作影像

Future Vision Housing — Housing as a Deal with Collective Intelligence

未来住宅—不同知识的聚合

DESIGNER: Dimitris Rotsios + George Bakoulis

Today's exposure to a continuous 'information attack' renders an individual a passive, alternating recipient of a vague, amorphous indefinability, upon which one is more and more dependent. This 'attack' is transforming 'values', traditionally associated to human nature, such as memory and knowledge into new ones, interwoven with information.

The environment is no longer what we knew it to be. Boundaries aren't absolutely visible or real. The boundary is the passage from a real world of direct perception to a 'field of relations' not always of material parts. In other words, the boundary can be grasped as the point of contact between the individual and the collective intelligence; the active and passive parts of an interaction, of which the habitation is the epicenter. The point that justly can downgrade 'fluid objects' to a 'household'. The point where all natural needs coexist with all the perceived experiences.

Indirect perceptions (TV, Internet) are 'external' and collective. Even in the 'net', digital recordings and software predetermine a range of possibilities, which is finite although seemingly endless. This totality is a simple indication not a speculation. Digital recordings are a potentialization and storage of possible texts and their projection a realization. In general, the display is a small window through which the user can explore the stock.

Direct perceptions are individual, internal and primary. Their evaluation is the user's "luggage" as well as a medium of memory, perceptual ability, imagination and thought; the active part of an interaction, the productive part of a world, which is characterized by random access.

[1] Those of direct perceptions. The most important are the kinesthetic ones, which are placed inside the house. Their transformation is evident in situations where they tend to contact the 'shell' due to the nature of their functions. For example, rambling, swimming, cleaning, eating, sleeping and reading.

[2] Those of indirect perceptions that are related to the 'potential world'. Their forms create a fluid space when contacting the 'shell'. For example, living, working, cooking, dressing, net-surfing.

The research of dipolar relations between memory and experience, knowledge and information, natural and artificial, produces an instant spatial expression. A realization of the changing of direct and indirect perceptions. The skin of structure is the medium - vehicle of the environment .An environment that emerges from the interweaving of real and potential. The result of an association. The epidermis through illustration is necessary for its definition. This new frame makes clear the necessity of the co-existence of natural and augmented reality. Euclidian space in the interior of the arrangement of uses corresponds to these directly interwoven with the concept of "luggage". This space is the "load" the user conveys, an area of primary production, a medium of memory and knowledge. The active part of an interaction with a collective intelligence. In this space are found uses of direct perceptions, with evident traces of their transformation in situations where they come or tend to come, in contact with the epidermis, because of their nature. Contact with the nature, contact with water, food, sleep is functions that belong to this category. The space that emerges between the above and the epidermis is space that favours interaction, contact with the collective intelligence.

大量且快速的信息，使个人存在变得模糊且难以定义，并且越来越依赖信息。与人性有关的价值观，如记忆、知识，也逐渐与信息混杂。

现存环境与我们所认知的已不同，界限不再存在，也看不到。界限连接我们可直接感受的真实世界与抽象的"连接的世界"，被视为个人与不同知识聚合的接触点。无论是正向或负向的互动，住宅均为其震中发生点，住宅会落实信息，此处居住者的需求与其感知的经验共存。

如电视、网络等间接经验是外在性的，多样折衷的，即便在此通讯网中，数码记录与软件已决定了许多可能性。但这些可能性是有限的，是暗示而非预期的，使用者可通过这些间接经验来探索这个世界。

直接经验则是个人的、第一手的，一般人会依使用者的记忆媒体、感受力、想像力与思想来评估。这也是互动中的主动部分，所以高生产率很难达到。

[1] 直接可感受的。这部分最重要的是运动型的动作，这些动作会依其性质而直接与房子有所关连，例如漫步、游泳、打扫、用餐、睡觉与阅读等等。

[2] 间接可感受的则是潜在的。一些其他的动作会塑造出一些流动的空间，例如生活、工作、烹饪、更衣和网络漫游。

AFFILIATION: Osarc.net
COUNTRY: Greece / 希腊

Interactivity in essence is the relation of accumulated knowledge with the collective intelligence as a contribution of the first to the reserves of the second, but also as a possibility of choices from the capabilities that second offers. The operating system of the above encounter.

Work, education, information are characterized by indirect perceptions, having a direct relation to the potential world and depending on their relation to the environment. A world it encompasses, without being purely "real".

A residence is a space that develops inside the epidermis as well as beyond it, transforming uses and positions. So each person, every organization, is urged, not only to receive something from a collective deposit, but also to suggest a view of the whole, a subjective structure.

即时性的空间产生于许多两极间的研究，即对记忆与经验、知识与信息、自然与人造物、产品与空间感以及对间接、直接感知的了解。结构其实是环境的媒介，环境又是真实世界与潜在世界的交织，可通过描绘来确定其定义。欧几里德式的空间其实会与使用者的负荷（luggage）相呼应。空间其实是使用者的负荷，是原始生产的领域及记忆与知识的媒介，是集体知识中主动的部分。在这些空间中我们便会发现直接感受的用处，发现依其性质转移的痕迹。在表层与上述功能之间的空间，其实是最有利于互动的，是与集体知识有所接触的。

在积累集体知识的过程中，互动本质与第一种感受相呼应，与第二种感受相悖，但它也是第二种感受的一种可能的选择。操作系统即依据上述关系而设计。

我们的工作、信息及教育因间接感受而有其独特性，它们既影响其周围的环境，又依赖于环境。这是它们所建构的世界，却并非全然真实。

住宅的发展在其内部也在其外部，不断地转换位置与用途。因此每个人、每个组织都是主观性的结构。[翻译　赵梦琳]

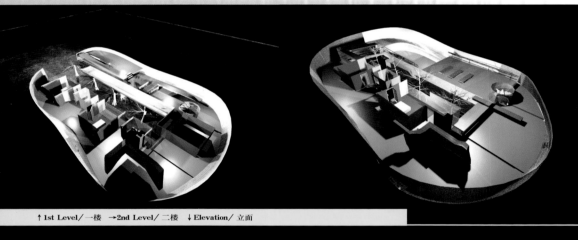

↑ 1st Level／一楼　→ 2nd Level／二楼　↓ Elevation／立面

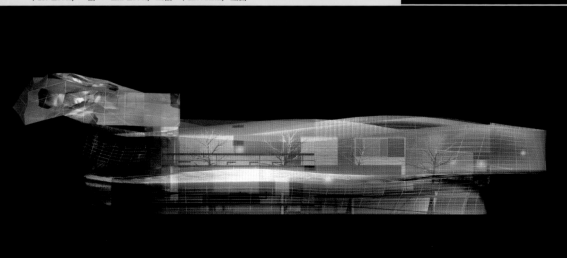

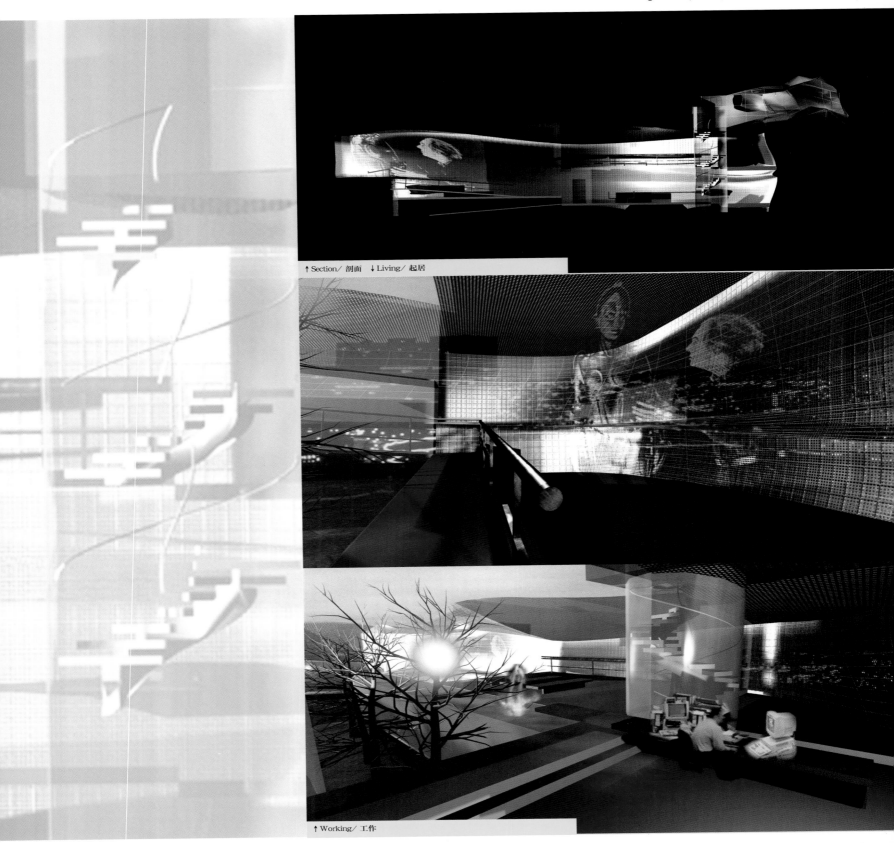

↑ Section／剖面　↓ Living／起居

↑ Working／工作

Slotmachine/ 吃角子老虎机

DESIGNER: Silvia Simoni + Lorenzo Noe + Luigi Fregoni + Marco Valentino

"Slotmachine" is an evolution of "Layers", a project presented to the International Venice Architecture Exhibition for the 7th Biennale.

In "Layers" components of the city are meshed to obtain a mix between relax and intimity spaces, speed and communication.

In " Slotmachine" causality creates relations between the complex of "Layers" and the world.

"Slotmachine" has four series of mood, which combined together determine the final situation.

Virtuality becames exploration of rendez-vous between architecture, landscape and emotions.

　　"吃角子老虎机" 其实是层次的演化，也是第七届威尼斯双年展的参展作品之一。

　　在这些层次里，城市会成为融合休闲、沟通、速度与交流的空间。

　　"吃角子老虎机" 其实有四种图案，放四种图案在一起才能呈现虚拟实境中探索建筑、景观及情绪间的关系。[翻译　赵梦琳]

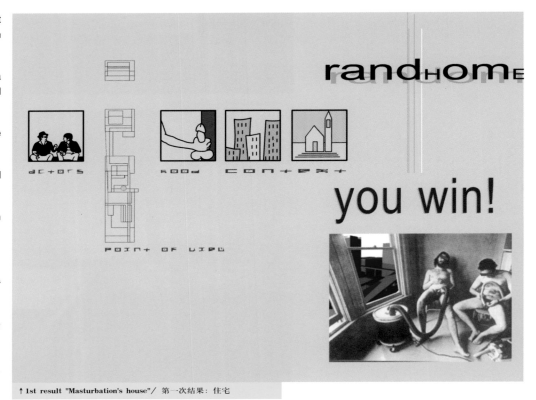

↑1st result "Masturbation's house"/ 第一次结果：住宅

AFFILIATION: @@@ Studio — Milano
COUNTRY: Italy/ 意大利

↑ 2nd result "I find my love in Portofino"／第二次结果：我在波多芬诺找到真爱

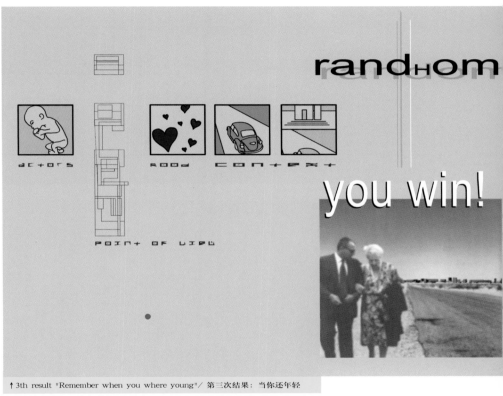

↑ 3th result "Remember when you where young"／第三次结果：当你还年轻

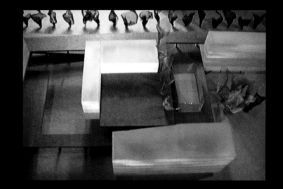

architectural
sioi machin

12

↓ Layers: Choose your home.../ 层次：选择你家   ↑ Layers: Combination/ 层次：结果

ON LINE

BIKE

STUDENTS

LIGHT
GREEN
NIGHT
WATER
WORK
SPORT
LEISURE
CULTURE
ART
SEX
MUSIC
FOOD
SHOPPING
FASHION

SHOWER
BATH
JACUZZI
WC
MIRROR
BED
SOFA
PLANTS
BOOKSHELVES
MODEM
TELEPHONE
COMPUTER
FRIDGE
DISHWASHER
WASHINGMACHINE
RADIO
STEREO
TV
TABLE
AMMOCK
BIKE
MOTORBIKE
BILLARD TABLE
TOYS
ANIMALS

SPORT

PARTY

WRITER

LEAD
GUITAR

TREES

BROTHERS
& SISTERS

HOME
all

Memory Constellations/ 记忆星群

DESIGNER: Peter Stec

Virtual space will increasingly transcend from physical reality simulation into environments without direct natural equivalents. The emphasis will be in structuring information, so that we may access it faster and more efficiently. This proposal should leave behind graphical or sculptural approach common in trees of knowledge or similar hierarchical structures of information. It develops methods of spatial self-organisation of information combined with automatic creation of a communication interface. In development, it is now only a mockup of the real engine and may look hierarchical because the evolution of data was stopped for the presentation. Upon visiting a web page, the engine assigns it a central graphical position in the 3D memory space. In that moment, a slightly turbulent repulsive movement emanating from the same center will affect it. Similarly to the galaxy drift, addresses will start to spread, with the logical consequence that pages visited first will be further away. Another process of organization starts after a first visit to a page. The more connections and traffic between two pages=addresses=information elements, the closer they get, mutual spatial attraction following mutual logical dependence.

The structures built in virtual space will represent personal association fields. As constellations, these will have spatial representations not necessarily dependent on spatial connections. Bounds will be established between discreet relating information entities, with the objective to automatically attain minimal information amount for a single statement, eliminating redundancies. Each information will beprovided with a spatial icon to which spatial connections will be made. This new spatial information architecture will provide an interface for web memories, using the experience of built architecture with up to date logical techniques and surprising outputs. For mnemotechnic reasons similar to the ancient discoursespasted into imaginary buildings for the sake of better remembrance, volumes will be generated around connections. These will also be dependent on the importance of a site and on traffic to and from other locations. Moreover, spatial traces of past development shall be left behind through connection points. The space does not have a privileged viewpoint. All the curves and surfaces were created with strictly spatial processes not connected to any top/front/side/perspective evaluation. They are described through forces, fields andevolution in time. All static representation is distorting. The result geometry should be complex & dynamic.

虚拟空间并不能真正取代真实空间，虚拟空间的重点其实是信息的集合，可让使用者快速且有效地取得信息。本方案旨在提供更高层级式的信息结构，创造沟通的界面。当使用者来到网站时，会启动一个 3D 的记忆空间，同时也会有一些干扰性的动作。如同银河一般，网址会逐渐散布。

早前浏览过的网页会逐渐远离。两个网址越靠近，越会从彼此的依存关系中获得空间上的吸引。本设计即希望以建筑提供网址上记忆的界面，以合理的技术发展出最令人惊喜的结果。在此建筑中并无预设的视点，所有的曲线及表面并非依透视来设计，而是以力、场所及时间演进来描述。任何静态的诠释都注定会被扭曲，最后呈现出的几何图形会是复杂且极富动感的。[翻译　赵梦琳]

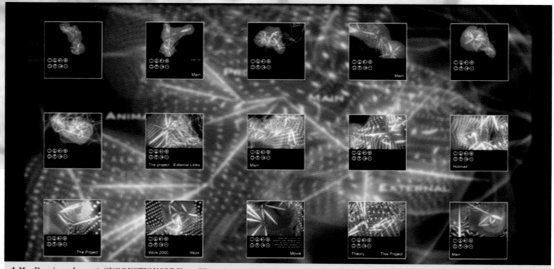

↑ My Previous Layout SUGGESTIONONLY: "Snapshots of a previous layout for print-as orientation" / 最早的建议案

AFFILIATION: Academy of Fine Arts, Vienna
COUNTRY: Slovakia / 斯洛伐克

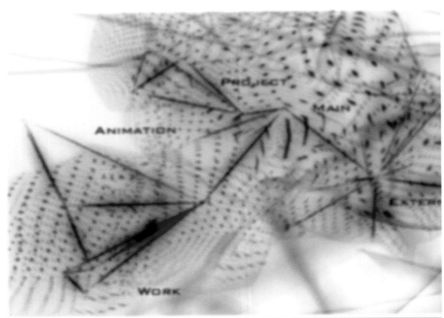

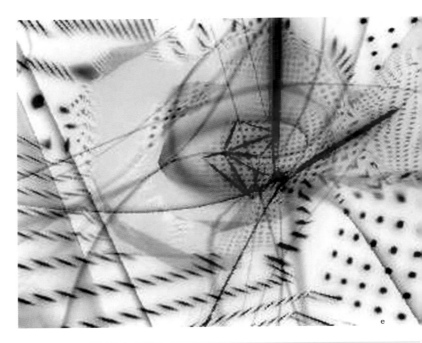

↑ Overall look of an organizing constellation "blurred, as background"／ 信息记忆星群的组织

a

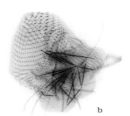

b

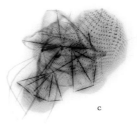

c

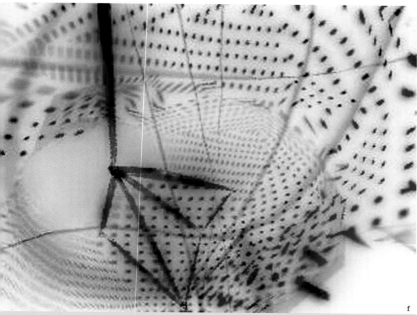 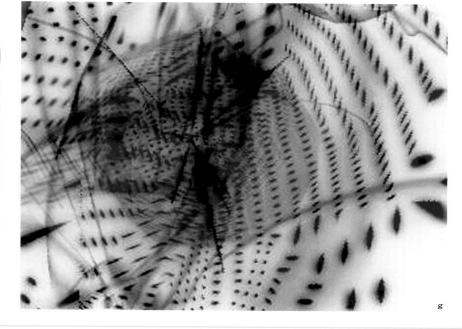

↑ e～g Interior spaces generated through associations／ 室内空间

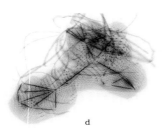

←a～d Process of self-organization of information "Please make the pics follow somehow — as filmstrip or so..."／ 信息之组合

Library of The Third Millenium/ 第三个千禧年的图书馆
DESIGNER: JMK Architects

"The millennium about to end has seen the birth and development of the modern languages of the West, and of the literatures that have explored the expressive, cognitive, and imaginative possibilities of these languages. It has also been the millennium of the book, in that it has seen the object we call a book take on the form now familiar to us. Perhaps it is a sign of our millennium's end that we frequently wonder what will happen to literature and books in the so-called postindustrial era of technology. I don't feel like indulging in this sort of speculation. My confidence in the future of literature consists in the knowledge that there are things that only literature can give us, by means specific to it." - Italo Calvino Six Memos for the Next Millennium.

Although he was referring to literature at the time, I believe Calvino's awareness of the future battles brought on by the digital revolution and his confidence in the qualities that would assure endurance through them has universal implications -- especially in the world of architecture, and particularly in this project. In our age of the World Wide Web and palm-top computers many have predicted the extinction of the library arguing that the gradual technological evaporation of the book necessitates the disappearance of the library with it. This oversimplified conclusion is a result of viewing the library simply as a container without any regard for its important societal and cultural roles. The irony of our recent global omnipresence is a new kind of digital seclusion with the computer terminal replacing our traditional social forums. The need for social condensers and symbols of shared cultural and community values is stronger than ever and, according to Elizabeth Diller, "it is architecture's role to resuscitate the authentic because of a general cultural loss of coordinates and technological saturation."

The role of a library is not only to store and organize our collective knowledge gathered over the last several thousand years but, just as importantly, to provide an environment in which that information can be absorbed and shared. This is an aspect of the library that will never change, regardless of how the nature of its contents do. The library is a place of comfort, repose, and minimal distraction. At the same time, it provides access to the best source of information -- other people.

Unlike the traditional books that fill our libraries today, the "books" of the digital library will not exist within bindings, but instead as nodes in a universal network of information. Besides containing the information within itself, each book also serves as a gateway or portal to an infinite number of related links. No longer confined and read the same way by everyone, front to back, books will instead be a field of information through which everyone chooses their own path and ultimate destination. Because it does not have to provide the storage space of a typical library, the digital library is for the first time built primarily for the people, not the books. This spatial freedom allows a complete rethinking of how one inhabits it. Digital tablets on wireless networks free one from the desk or table and its stack of books. Your physical space, just like your virtual space, is defined by how you choose to navigate it.

Built on the former site of Berlin's spectacular Anhalter Bahnhof, a 19th Century rail station severely damaged during the Second World War, the Anhalter Digitalbibliothek is a library for the next millennium. Once a terminus for the vast European rail network, the site is transformed into a point of entry into the global digital network while providing an important public forum for learning and social interaction. Taking cues from both its history and the technological possibilities of the future it provides one solution of how our physical space may be inhabited like our virtual space.

"过渡到新的千禧年，会产生新语言以及更富表现力，更具想像力的新文学，对书籍也会产生影响。我们称之为后工业时代。但我仍然对文学有信心，因为文学会给我们真正想要的。"——卡尔维诺：《给下一个千禧年的备忘录》

卡尔维诺虽然指的是文学，但也可应用于建筑。网络与PDA的盛行，可能会使图书馆消失，这样的看法其实是把图书馆过度简化了，忽略其社会、文化的角色。

图书馆并不只是储藏、组织知识的场所，它其实提供了一个可吸收，分享信息的环境，不管信息的形式如何，这点是永远不会变的。图书馆必须舒适、宁静、能让人专注，随时可提供最佳的信息。

数码图书馆的藏书不会只是建筑物内的书本，而会是一个信息网的节点，每本书都是连接更多信息的门户，阅读也不会再有次序，读者可自行选择需要的段落。由于数码图书馆不再需要提供藏书的空间，它的空间因此更自由，使用者可自行选择其漫游方式。

我们选择柏林19世纪的火车站作为数码图书馆的地点。

火车站昔日是欧洲网络的节点，现在则是信息学习与社交的中心。以历史为依据，我们尽力探索科技的可能性，让真实空间与虚拟空间合二为一。[翻译 赵梦琳]

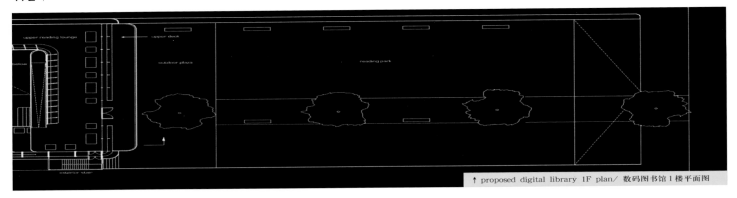

↑ proposed digital library 1F plan／数码图书馆 1 楼平面图

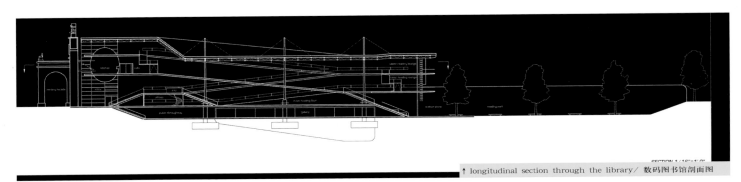

↑ longitudinal section through the library／数码图书馆剖面图

↑ The multi-story atrium as the reading room／挑高阅读空间

Liberal State/ 解放状态

DESIGNER: Yi-yen Wu/ 吴宜晏

The cyberspace is now having an extensive influence on our practical life. The large volume of duplicity, and interactive and rapid transmission are impacting the way how people get known with the world. It changes so fast that it produces uncontrolled and uneasy feelings.

Design Concepts   "Liberal State " is what the author felt during the period when staying in Wu-Shan-To. "Liberal " means that the object exists respectively without merging with others or that it separates from the matrix. Applied in the architect creation, it is a kind of resistance in inner heart, developing from the drastic shift of the outer space, trying to explicate the inner state of the heart. It is a kind of uneasy and unstable state.

Hereby I try to use "Liberal State" as a basic concept, through the transformation and examination of object, search for the element in architect space and creation to proceed the construction of my own characteristic. Therefore, I deem the operation as a test, starting with photography, to build my own concept of aesthetic space.

Model
Technical Tools: Plastic roofing board, Steel wire, Wire rope
Size : 300cm × 500cm × 250cm
Date of the Work: 2000

What "Liberal State". In my work I make an attempt to exploit concrete substance (plastic roofing board), suspending in the open space, yielding the effect of float and instability, to reflect the relation between real and virtual world. Then I utilize its penetrating feature of this kind of material to form a void atmosphere.

"Liberal State" is a kind of author's spiritual reaction to the distrust and instability of the virtual world when he lives in the real world.

Drawing   "Liberal State" is the escape from the world and incompatible state after resistance. It produces floating and separating relation between the matrix and outer world.

虚拟空间正对我们实际的生活产生深远的影响，大量的重复、相互作用及信息的迅速传播，冲击着我们对世界的认知，瞬息万变的虚拟世界引发我们心中的不安。

"解放状态"是生活在现实世界的作者的心灵对虚拟世界所反映出的猜疑与不安。

设计概念   "解放状态"是作者暂住乌山头时所感受到的，它意指物体的个体存在，不与外界相融合或指与母体脱离的状态。应用于建筑创作上，则是由于外在环境的激烈变换而产生，并试图解析心灵内在状况的一种压抑，是不安并且不稳定的状态。

我试图以"解放状态"作为基本概念，通过对于物体的检视与变化，进而找寻并着手构筑本身特质所赖以实现的建筑空间元素。所以我将此次的操作视为一次实验，由摄影出发以建构我对于空间美学的审美观点。

模型
技术工具：塑胶屋顶板，不锈钢线，线绳
规格：300cm × 500cm × 250cm
创作日期：2000 年

作品试图以悬浮在开放的空间中的实体物质(塑胶屋顶板)所产生飘浮不稳定的效果，反映现实与虚拟世界的关系，接着利用其穿透特性形成虚空的气氛。

描绘   "解放状态"是逃离现实世界后的矛盾状态，它造成母体与外界的悬浮分离。

[翻译 蔡咏岚]

AFFILIATION: Graduate Institute of Building Art, Tainan National College of Arts/ 中国台南艺术学院建筑研究所

↑ Photo of Space One／ 空间一

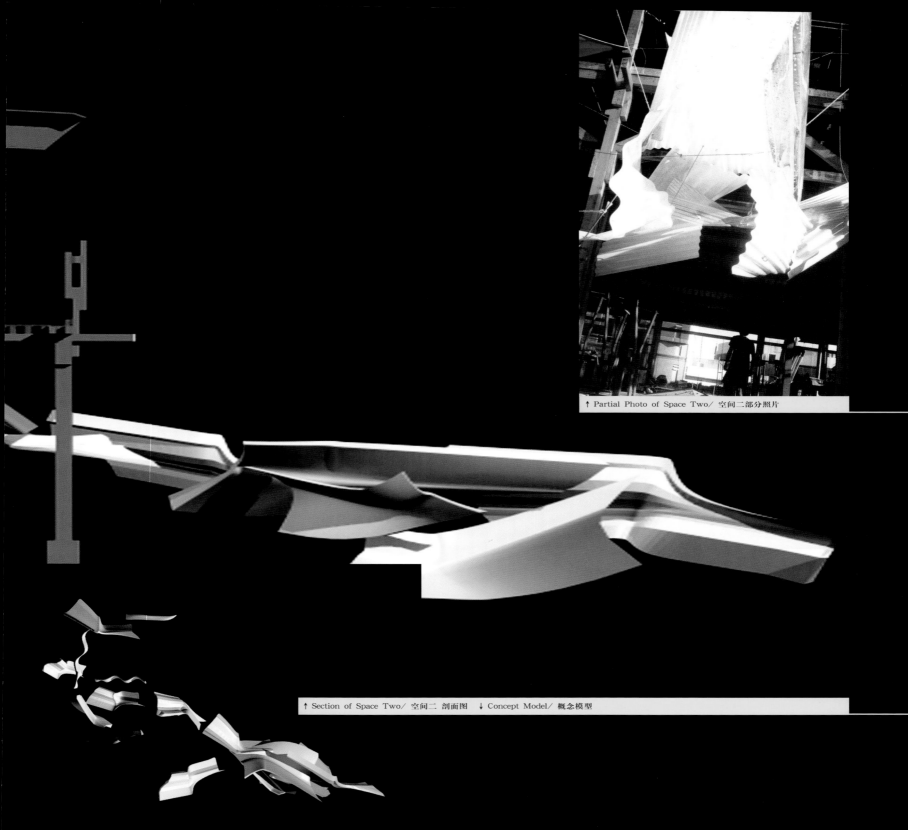

↑ Partial Photo of Space Two／ 空间二部分照片

↑ Section of Space Two／ 空间二 剖面图   ↓ Concept Model／ 概念模型

Wylie Associates Web Site/ 威利建筑网站

DESIGNER: David Wylie

Digital Design Approach   Communicating dynamic, cost effective architectural solutions in a digital world, this is the aim of WYLIE ASSOCIATES and its architecture on the World Wide Web.

www.wylie-associates.com is the company digital architecture, its presence on the web and WYLIE ASSOCIATES submission for the FEIDAD award. It is not a static architecture - it evolves and adapts to the company needs, the demands of its visitors and a world of changing technology. Created as a digital 3-dimensional environment, it is not designed to be built in reality it stands for what it is: conceptual space for abstract ideas routed in the real world. It is literally the architecture of the new economy.

WYLIE ASSOCIATES and Digital Services   Founded in 1997 as an architectural practice registered with the Royal Institute of British Architects in the UK, WYLIE ASSOCIATES now operates in realms as diverse as communications, resourcing and digital architecture.

WYLIE ASSOCIATES meets people needs by exploring building design concepts in both the real and virtual worlds exploiting opportunities for clients offered by the World Wide Web bringing the digital world to architecture and architecture to the digital world.

WYLIE ASSOCIATES and Digital Architecture Design on the World Wide Web   With hits numbering tens of thousands a month, www.wylie-associates.com is regarded as a world knowledge base on the complex topic of intelligent buildings an architecture that adds value to the built environment through embracing leading-edge concepts in design and the digital economy.

www.wylie-associates.com demonstrates WYLIE ASSOCIATES approach to a global audience. Visitors identify with its 3-dimensional interface, allowing a set of complex ideas to be laid out clearly and imaginatively. Realising that people come to the web site for research, networking and for the services the company has to offer, WYLIE ASSOCIATES designed a digital

architecture that meets these different needs. As a virtual environment for the web, the gallery, public library and private library allow visitors to explore a knowledge-base of projects, outline summaries and in-depth articles on WYLIE ASSOCIATES work. Visitors can move around freely from one cyberspace to another learning from others and communicating new ideas in the process.

It is the Virtual Office at the core of the web site however that is central to WYLIE ASSOCIATES digital architecture. From here, visitors gain an insight into the company service philosophy and clients have access to the inner workings of their projects by engaging dynamically with project-linked on-line databases. By continuing to expand its uses, digital architecture is becoming an essential tool in enabling WYLIE ASSOCIATES clients to meet their aspirations. The company digital presence reaches out to its projects in the global market. www.wylie-associates.com is WYLIE ASSOCIATE link between its real-world architecture and that which crosses boundaries of time and space to service a new type of leading-edge client in a dynamic knowledge economy.

AFFILIATION: Wylie Associate
COUNTRY: U. K. / 英国

威利建筑网站旨在建立更有效率的建筑设计。 虽为一3D模型网站，呈现的却是灵感源自真实世界内的抽象空间，并非真要建造，它是新经济时代的建筑。网站建立于1997年，重点经营项目是传送、建立资料库及数码建筑。

网站中收集了虚拟世界及真实世界的设计，以满足客户需求，融合了建筑与数码世界。

网站上每个月有数以万计的设计方案，均以最新科技来反映数码时代的建筑需求，它的对象囊括全球，3D界面容许上网者表达最开放的想像力。

网站上的数码建筑，可满足客户研究联系等需求，包括对作品的介绍，其服务功能包括图书馆及画廊，上网者可在网站间自由漫游，交换心得。网站的重点空间为虚拟办公室，上网者可由此对网站及其设计方案有更深入的了解。数码建筑能满足客户日益增多的需求，并将触角伸向全球。

此网站超越时空，并在知识经济时代抓住了最顶尖的客户。[翻译　赵梦琳]

↑ Overview of the on-line 3D environment of www.wylie-associates. com/ 网站全景　→Wylie Associates Virtual Office entrance/ 虚拟办公室入口

wylie associates

total networked services:

websites
graphics
databases
architecture
3D visualisation
communications
intelligent buildings
resource management

www.wylie-associates.com

↑ Wylie Associates' Project Gallery／ 画廊
↓ The Orientation Space with views of the Offices, Subterranean Maze, Libraries and Gallery／ 由场地广场看办公
室、地下迷宫、图书馆及画廊

eggNOMA — Dive into Your Own World/ 投入你的个人世界

DESIGNER: Michael (Benarroch) + Mehdi (Berrada)
Members of Po. D/ Experimental architecture & design association based in Paris

Our research led us to react within the globalisation phenomenon, which takes place and spreads out nowadays, leading to a loss of human identity. This proposal is a punctual answer to a need created by a new race of urban nomads.

It is a new way of living the city within a short time, which each one can find his appropriate measure. This short time repeated at different intervals reproduces the housing.

An individual graft allows each user to set the internal space of a bubble, referring to memorized atmospheres, body positions, images and sounds stored in the graft.

eggNOMA is then the wearer which allows transferring memories, marks and identities within a space, thanks to an hemispherical screen and to an ergonomic membrane, powered by inflatable balloons.

These atmospheres and memories can be exported or imported as files and a link can be established between bubble and web users around the world. It then becomes possible to recreate a specific environment to each activity and proper to each individual.

Therefore, we are able to find again or to create the atmosphere of our own office, the ideal game or cyberspace interface, but also a therapeutic environment, to meditate and to relax for example, all together in a visual and ergonomic way.

此设计旨在研究如何对全球化的现象做出反应，尤其全球化常使人们丧失自我定位。此设计亦针对都市里的新游牧族，它也是城市内一种新的短期生活方式，每个人都可找到正确的方法，此种短期生活的重覆即为住宅。

此设计容许每个使用者在泡泡内设置自己的空间，包括气氛、身体位置，影像及声音。

eggNOMA 则可以通过 eggNOMA 薄膜上半球形的银幕转化空间内的记忆、印迹及特征，此银幕由汽球驱动。

这些记忆与环境，可如档案般进出，全球的网络使用者间也可有所连接，如此可针对不同使用者的需要来营造特定的氛围。

我们也可针对自己的办公室塑造所需的气氛，模拟理想游戏的虚拟界面，可沉思或放松的治疗性环境，这些都是可见的且符合人体工学的。[翻译　赵梦琳]

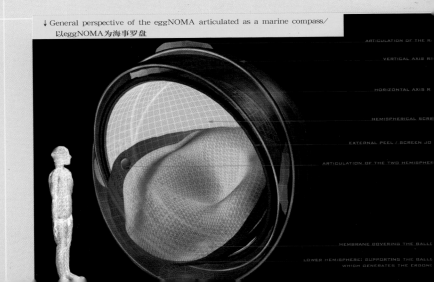

↓ General perspective of the eggNOMA articulated as a marine compass/
以eggNOMA为海事罗盘

AFFILIATION: eggNOMA
COUNTRY: Morocco, Israel/ 摩洛哥，以色列

A Virtual Gallery / 数码画廊

DESIGNER: Dace A. Campbell, AIA

Virtual architecture is the spatial expression of our digital society, liberated from the bonds of the physical universe and given new life in its freedom from physical constraints. Having no physical expression, virtual architecture exists exclusively in cyberspace, accessible from the Internet via computer interfaces. Virtual architecture uses architecture as a metaphor to create a meaningful place in the context of abstract information-space.

This project explores the limits of this metaphor, to gain an understanding of how the principles borrowed from the design of physical space are appropriately applied to the virtual realm. It is a virtual gallery, exemplifying what is possible in the design of virtual environments using this spatial metaphor. The gallery features a collection of virtual environments created at a virtual reality research laboratory, and is intended to organize and to demonstrate those virtual worlds. The virtual gallery was designed to be experienced in real-time with virtual reality hardware, software, and interfaces. It is available here for your exploration as a collection of images captured from such real-time walkthroughs.

ENTRY The virtual gallery is expressed on the "exterior" as several interlocking forms that create a sculptural composition of "interior" spaces. There is a single entry point into the virtual environment, at a vestibule, which offers a view towards the gallery.

CIRCULATION A circulation spine leads visitors from the entry vestibule into the gallery spaces. Visitors follow abstract signage and interpret directional cues from forms in the circulation spine to orient and move into the galleries.

GALLERIES Four galleries are arranged around the spine with unique proportions and orientations to convey that they serve a special function. Each is a different color to aid orientation, with the warmest colors nearest the entry. Each gallery features a display wall with images representing virtual worlds and serving as hyperlinks to those worlds. The time it takes to load the environment is represented spatially as a tunnel one moves through until the new world is loaded.

MAIN HALL The main hall at the end of the circulation spine is a gathering place for visitors to meet and interact. The hall is created with the flexibility to add new functions as collaborative interaction among on-line participants increases in virtual space. The circulation spine passes through the main hall, ending in an enframed view of the void and enhancing the awareness that the gallery has no physical context. Colorful corridors extend out in all directions from the hall, expressing links leading to other virtual environments beyond the gallery.

ARCHIVES The archives support the galleries, storing and displaying virtual environments that are no longer actively featured. Its orientation and proportions differ from the circulation spine and the galleries, expressing its unique function. Dim lighting is used to de-emphasize the entry to the archives to the casual visitor; additional lights turn on upon entry into the archive to aid the operation of curators and navigation by purposeful visitors.

数码建筑形象地表达了我们身处的数码社会，它赋予我们更多环境上的自由，虚拟空间其实只存在于电脑的界面中，就虚拟建筑而言，其实质是在抽象的信息环境中创造一个有意义的场所。

本方案旨在探索虚拟空间与真实空间的关系，真实空间的设计原则如何适用于虚拟空间，如何应用于虚拟画廊中。此画廊集合以虚拟实验室创造的虚拟环境，亦可自行组织及展示这些环境。

入口 画廊外观以室内空间为依据，呈现如雕刻般的效果。

动线 室内空间自然形成一动线，暗示访客自行进入画廊空间。

室内 画廊包括四个单元，沿主要动线组织而成，每个单元色彩计划均不同，以帮助使用者定位，越靠近入口色彩越暖。

大厅 大厅主要是供访客碰面与互动。资料库用以保存画廊内暂时无须展示的资料。

[翻译 赵梦琳]

→View of the main hall, including the enframed view of the void beyond./ 大厅内景观

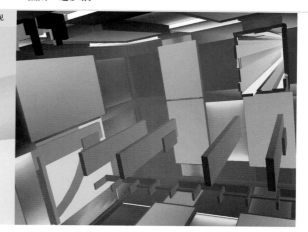

AFFILIATION: NBBJ and The UW Human Interface Technology Laboratory
COUNTRY: U.S.A./ 美国

The Gate/ 闸　门

DESIGNER: Han Chung + George Huang + Andrew Wong

As information penetrates the Internet, there needs to be a place where individuals can be guided.

The Gate is a portal that steps across the barrier between the physical world and the Internet. It is an exhibition center where information can be obtained; interaction between the physical and the virtual is achieved, and a guiding direction to the tech world is given. In concept, the building is floating between the above-ground, which symbolizes the physical world and under-ground which represents the virtual world. The building contains an information forum, an exhibition hall, information tube throughout the building, and an exhibition ramp.

信息泛滥于网络，人们反而更需要安身立命的场所。

此闸门介于真实世界与网络之间。在此展示空间内，人们可获得互动信息，飘浮于地面的建筑物象征真实世界，地面下的则象征网络世界。建筑物本身包括信息中心、展览厅、贯穿建筑物的信息通道与展示坡道。

[翻译　赵梦琳]

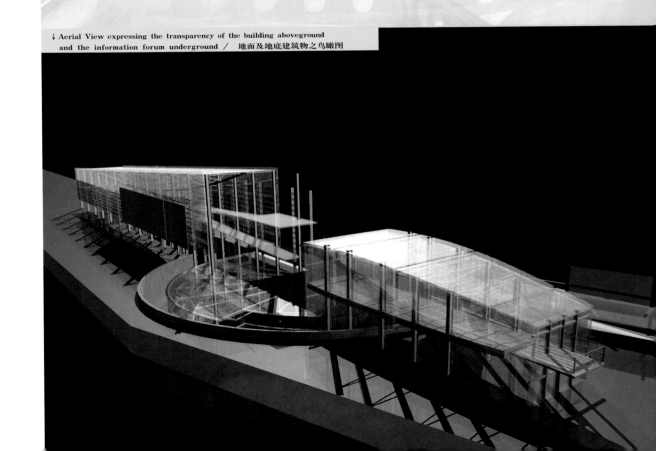

↓ Aerial View expressing the transparency of the building aboveground and the information forum underground ／　地面及地底建筑物之鸟瞰图

COUNTRY: U. S. A./ 美国

TransGen(Etic)Structure for Ex Yugoslavia — Amniotic Space/
给前南斯拉夫的遗传构造—羊膜空间

DESIGNER: Juricic Bruno + Franinovic Karmen

[1] What if no road comes to house? What's happening when our conceptual view is fixing the words as "hospitality, proximity, stranger"?

[2] For a moment we perceive this words: the "Other-ness", the"Strangness", but our Habitus of Thought and memory assimilate it into its familiare landscapes of the question "Where come you from? That is the negation of Hospitality and is not permitting the seeting up of connection Xeno-Xenia/ Stranger-Alliance. So we have to force Thought to the Out-field ,to break throught the lines that delimit our conceptual horizons and to produce the "Third-Room"the Amniotic Space, the angel-angle machines that affect the organization of dwelling, and the construction of identities ,which in turn shape the way we act ;the question of our collective and individual positionig and placement and that can reconfigure notions such as private, comunity, efficiency.It's about constructing the Beat-Off, the TransGen(Etic) structures in the convergent sequences of relation territories/ identities/ buildings/ spaces/ people.

[3] TGE structures break down the idea of the exisiting logic of the three cartesian axis as a semantic cage, by absorbing it in a further transformation and construction of AMNIOlogic. They form an fluid aggregate but not a totality. The surface-product is not a simbiont wich fills the voids; it behaves as an AMbiont moving from cartesian logic to construct the logic of hospiality. It cannot be defined clearly because it overflows and transform, rises there where matter and information flow simultaneously, it's bubbling and gurgling, oscillating wildly in the gaps of the actual structure of the city. It can be affected and can create consistence,a new space continuum determined by technological implementations which increase its dimensional features, which isn't a product of building industries, but a combination of aircraft and automobile industry, fashion and bio-logical analyses.

[1] 若我们前方无路可供选择，当我们的概念视野专注于"好客、邻近、异客"时将会发生什么事？

[2] 我们暂时意识到了"他人"与"疏离"，但我们思考与记忆的习性将之吸收进"你从哪儿来？"这个熟悉的疑问之中。这是对好客性的否定而且不允许"陌生—好客"和"异己—联盟"的连接。所以我们要进行超领域的思考，穿透原本思想视野的界限并且制造"第三房间"羊膜空间，影响居住组织的"天使—角度机器"来塑造我们行为的身分；集体与个别的定位重新定义了私密、社群和效率等概念。这关于"节奏—关闭"的建构，在领域、身分、建筑、空间、人的关系的收敛序列中的遗传构造。

[3] 遗传构造瓦解了现存笛卡儿逻辑的中心为一个语义的牢笼，并将它进一步吸收转换成为羊膜空间。它们形成流动的集和但并非总和。表面产物不是填满虚空间的花蕊；它作为由笛卡儿逻辑转变为理解力逻辑的双向物件。因其满溢与变化而不可能将之清楚定义，在物质与信息随机流动之处升起，汩汩地流，冒泡，在都市实际的构造裂口中狂野地摆荡。它可被影响并制造一致性，增加了由向度特性的科技成就定义的新空间的连续，它并不是建筑工业的产物，而是航空工业、汽车工业、时尚与生物学分析的合成。[翻译 蔡咏岚]

↑ Amniotic Space   Within the context of place, space and a structure is a kind of noice or chaos known as multiplicity. A multiplicity is a source of transformation composed of elastic sufaces configured into indiscernibile trajectories; and with it's mobility as a movement of indeterminancy creates consistency. The amniotic surface protect us in order to liberate and to create the space of hospitality. It is a maternal cavity wich produce hospitality and offers a space for it's production. It affect the construction of identities. / 羊膜空间

在空间涵构中，空间与构造被视为多样性的噪音或混乱。多样性是变化的根源，由弹性的表面组成难以识别的轨道;机动的特性作为不确定的运动而创造了协调性。羊膜的表面解放并创造空间的好客性而且保护我们，它是种母性的腔穴制造并提供好客所需的空间，影响了身份特性的建构。

AFFILIATION: Istituto universitario di architettura di Venezia
COUNTRY: Croatia/ 克罗地亚

Future Architecture/ 未来建筑

DESIGNER: Richard Koeck, Dipl.-Ing.(FH)

Architecture and the life in the city of the third millenium might be one in between virtual realities and our real world. We might find our physical bodies in an environment surrounded by more than one reality. Inhabitants of cities might live as much in them on a virtual basis as they live physical in them. People might have the ability to create their own virtual cities according to their individual human needs and desires. A significant amount of their time they might spend physically isolated, but brain-stimulated in a virtual world - in virtual city. Image "zoom 1" shows a city as we know it today. A multi-layered traffic dominated vital city, melting pot of cultures and social life. The upper left corner shows the binary code in mirrored form, indicating that the shown image is view from the inside of a computer monitor into the outside world. Image "zoom 2" shows how a computer system zooms closer to one of the buildings, looking for the "one reality" of its own existence. The binary code symbolizes the dominance of computer technology is projected one the architectural facades. Image "zoom 3" shows how a computer system zooms closer to one of the facades, recognizing that there is apparently "more than one reality", projected one the building facade. One projected image on the facade shows exactly the previous image "zoom 2". Image "zoom 4" moves through the facade into the building, finding a child as animator and creator of the virtual city reality, shown in the previous images. The child's real physical space is rather empty and cold and is spatial constricted by a large computer in the background. The city of the future might be by far more detached from materiality as we can imagine today.

在下一个千禧年内，建筑和城市生活会存在于虚拟世界及真实世界之间。我们的存在，会被不只一个实体所包围，待在真实世界中的城市居民和生活在虚拟世界中的一样多。市民也会依个人需要来创造自己的虚拟城市，在许多时候他们在形体上会是孤单的，但在精神上却与许多人连接。Zoom1 是我们现在的城市，不同层次的交通主宰了城市的文化融合及社交生活，在左上角以镜射的方式来表达双生现象（二进制），其实是将电脑屏幕内部投射至外界。由Zoom2可看出电脑系统如何与建筑物接近，找出其存在的惟一的真实，建筑立面也可达到电脑的二进制。由Zoom3可看出当电脑较靠近立面时，会比真实还接近真实。Zoom4投射在建筑物上的图像会一直延伸至室内，而且会出现空间的创造者：一名儿童。这名儿童的空间是空洞冷漠的，并会为背景中的电脑限制。未来的城市将脱离现实的限制。[翻译 赵梦琳]

↓ zoom 1, zoom 2, zoom 3, zoom 4　缩图 1、缩图 2，缩图 3，缩图 4

ZOOM 1 - INSIDE VIRTUAL REALITY

ZOOM2

ZOOM 3 - INSIDE VIRTUAL REALITY

ZOOM 4 - OUTSIDE BINARY REALITY

AFFILIATION: Visiting Assistant Professor, College of Architecture, University of Oklahoma, USA
COUNTRY: Germany/ 德国

Internetmuseum MMKI / 网络美术馆

DESIGNER: Andreas Korte (design and concepts) + Zeitraumsysteme GmbH, (technical support)

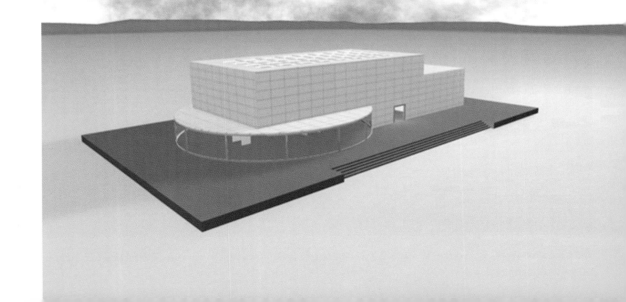

The MMKI (Museum fuer Modern Kunst in Internet) is an innovative platform for contemporary art in the internet. It offers generous space for a variety of selected exhibitions and projects. The visitor moves with simple drag & pull of his mouse through the rooms of the museum, fully interactive and in real time. As the museum is constructed with a new 3D-software it can be viewed without any plug-in. Additionally to the current exhibitions the museum offers information about the artists and their works, as well as a documentation of previous projects. Exhibitions in the MMKI can be visited from all over the world, 24 hours a day. The MMKI is the result of the cooperation between Andreas Korte (Duesseldorf,Germany, design and concepts) and Zeitraum systeme GmbH, (Essen, Germany,technical support).

网络美术馆旨在容纳当代的网络艺术，容纳各式展品。使用者只需点击鼠标，即可于馆内遨游，有充分互动的机会。除了作品，美术馆也介绍艺术家及其相关作品，无论白天或晚上全世界的人均可造访此美术馆。[翻译　赵梦琳]

Exhibitions:
February   Opening show by Andreas Korte, Duesseldorf 2001
March   Christine Ehrhard, Duesseldorf
April   Christian Freudenberger, Duesseldorf
May   Lothar Goetz, London

Special Events: The MMKI was presented in the "Grosse Muenchner Kunst- ausstellung", Haus der Kunst, Muenchen, 16.11.2000-27.1.2001

↓ View from outside／外观

COUNTRY: Germany / 德国

The Interactive Virtual Cafe Shop / 虚拟互动咖啡馆

DESIGNER: Te Lai/ 赖德

The TV wall, that is a feature of the electronic culture in city and disappearance of the place to exist elsewhere. We live in the urban space, which full of media signs and the simulation of our experience. When we are into the Internet space, however, "I" has built many identities that comes out with the Internet and We own ourselves' identity, the Internet makes the person to have a more private space, getting into our life and the real world parallelly.

The interactive essence is not a real space experience, that multiplying the participation in the cyberspace. The space and the object is the indefinite state, it is moveable and regenerate to blur the public and the private domain at the real existence to express a continuously movements and it's constantly collapsing and recombination on the digital architecture place of the cyberspace with us everyday. The process which exists the objective and the objective collide to each other and the relationship of separation. It moves continually to change the order from the interactive movement random state and to make the space, creating a new form of this site to resemble the non-order condition of the internal history.

Every object appears to the nonentity of the normal experience to be conscious, trying to experience from perception and reception of the space. The fragments of the objects are the reproductive identity of the space, avatar to mean the indicate, real space image and user's identity .

The Avatar in the Internet is a translation sign from the real world identity. The main body of human being is getting faded away. It turns into the media of simulation to superimpose on the object. The character reproduces the image of the real world identity.

That is an experiment of architecture, the cafe shop attempts to combine and connect the real and the virtual client, study identity and avatar and to experience the space bridge of interface in digital architecture.

电视墙是都市电子文化与场所消失的一个特征。我们生活在充斥着媒体设计与感官刺激的都市。而当我们进入虚拟空间中，"自我"随网络的出现建构出许多个身分，我们拥有自己的身分，网络提供个人更私密的空间，构成平行的现实生活与网络生活。

互动的本质并非现实的空间经验，而是使发生于虚拟空间的参与倍增。空间与物体处于未定状态，可移动与再生的特性模糊了公共与私人的领域，实存表达了数码建筑虚拟场所的持续运动、瓦解与重组。目标与过程相互碰撞分离。它持续移动改变随机互动运作的顺序，创造空间，创造出新形态来表现无秩序的内在历史状态。

就常规经验而言，每件物体都显得是有知觉的，试图以感知与反应来体验空间。物体的构成是空间再生的身分，体现所指出的观点，是现实影像与使用者的身分。

网络上虚拟互动地呈现一个人是现实身分的转化标志。人类的主体正消逝枯萎。它变成叠加在物体上的媒体刺激。角色再造现实身分的影像。

它是建筑的实验，咖啡馆试图结合连接现实与虚拟的客户，研究身分与体现，体验数码建筑中介面的空间桥梁。[翻译　蔡咏岚]

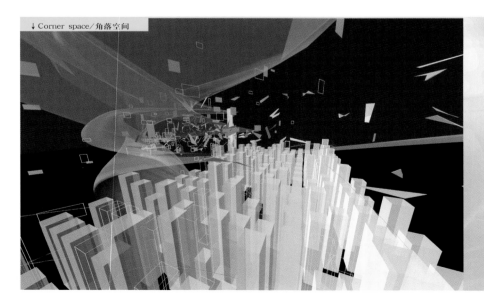
↓ Corner space/角落空间

AFFILIATION: Graduate School of Applied Arts National Chiao Tung University/ 中国台湾交通大学应用艺术研究所

Digital Cinematheque — between to See and to been Seen
数码实验电影院—看与被看

DESIGNER: Lin Ying-Tzu/ 林盈孜

"Digital Cinematheque" is a virtual theater displaying different kinds of arts and performances. Any type of performance is ademonstration of relationship between to see and to be seen. It is undeniable that the forming of the theater in ancient Greece isbased on how Greeks viewed the relationship between the performers and audiences. The shape shows that physical conditions, such as acoustic and visual, have been seriously taken into consideration.

As in the background of VEs nowadays, though the physical conditions differ, the essence of performing space is still the same. The theater experience is an entertainment activity. Entertainment is message, audiences digesting messages, while actors, singers, and dancers producing the messages. Theater is responsible for disseminating the messages and information. So, the essential design concept is "How does this virtual theater, Digital Cinematheque, take shape from its spatial essence? ", and " How are the messages transmitted through the relationship between to see and to be seen?"

Preliminary operation in [Part 1- Painting] explores how information determines the composition of a space. The walls in this design demonstrate different ways in which people recognize spaces and perceive environments. Space is shaped through the display of art works, representation of the painting process, and re-creation of art. [Part 2- Dancing] shows that when the visible walls in Part 1 disappear, how the interface, the invisible wall between the stage and audience, influences the way user understands the performance and surrounded environment. The interaction between users and VEs is highlighted. [Part 3-Chinese Poetry] further discusses issues of multimedia environments and metaphors of forms by the demonstrations of different approaches towards the presentation of a Chinese poem. Exactly the same poem is shown in different forms, such as text, codes, images, music, and vocal recital included. The spirit of the project is to provide windows for looking out, behind which are diverse information and messages, the true face of VEs. In which manner, VE becomes a programmable place where participants can interact with it throughout the design interfaces.

The purpose of this proposal may not to present a complete project, but to demonstrate a possible way toward the studies of virtual environmental design.

数码实验电影院是个展示各种艺术与表演的虚拟电影院。任何一种表演都是看与被看的关系。无可否认，古希腊戏院的发展的基于演员与观众之间关系，其形状显示了物理条件，例如视觉与听觉，是经过谨慎考虑的。

虽然在电子化背景的今天，物理条件改变了，但是表演空间的本质未变。剧院的体验是一种娱乐活动，而娱乐是一种讯息，观众接收由表演者、歌手、舞者发出的讯息，剧院负责传播信息与资讯。所以设计概念的关键在于"这个虚拟剧院怎么从自身的空间本质成形？"，以及"信息怎么从看与被看的关系中传达出去？"

[Part 1—绘画]的初步操作是为了探索信息如何定义空间的性质。这个设计里的墙表明了人们识别空间与感知环境的各种不同方式，空间通过代表创作过程与艺术再造的艺术品的展示而成形。[Part 2—舞蹈]显示当Part1可见的墙消失后，舞台与观众之间隐形的墙，影响使用者了解演出和环境的方式。使用者与VEs间的交互作用被突显出来。[Part 3— 中文诗歌]通过中文诗歌各个不同的呈现方式进一步讨论多媒体环境和形的隐喻。同一首诗以文字、符号、影像、音乐、朗读等形式呈现。本方案的精神在于提供向外观看的窗，在此之后是VEs的真面目：各种互异的资讯与信息。VE成了一个参与者在各个设计的界面中可以互动，可设计操控的场所。本方案的目的并非为了呈现一个完整的方案，而是为示范设计虚拟环境的研究方法。[翻译 蔡咏岚]

←The forming of space in [Part 1-Painting] springs from the visual perception and the way individual recognizes things./ [Part1-Painting]中空间的形成是来自个体对物体视觉感知与辨识。

AFFILIATION: Cheng Kung University, the Department of Architecture/ 中国台湾成功大学建筑系

Earthquake/ 地　震

DESIGNER: Angelo Micheli (project leader)+Alberto Bianchi
+ Claudio Venerucci(arch employee)+ Elena Riva(graphic designer employee)

"new millenium city"

The architectonic earthquake will be the beginning of a new dynamic relationship between peoples. Perhaps the new city will forget about the history of the past and will start, with architectonic easiness, building technological huts in order to continue on new distances to the completion of the city system. The more and more sophisticated technology will be the line guides of the development of the city, the new path of development and the new behaviors will be characterized from technology but, the isolation of human relation will have to be avoided. The modularity of the city will become more and more controlled in order to allow to accelerate every relationship but, the city will revive after having found again the patience and after having accepted that eagerness is the greatest limit of speed. The facility of penetration between the lines that separate places, will bring back people to living freely the new open cities and to use the technologies only as a support in order to proceed in the acquaintance. The new city will be endowed with intelligent trees allowing the imediate connection to all technological plagues. Technological complexity, will stretch to the rationalization of every thing, constructing architecture will endure this formulation. The diversity of the architectonic language will be given, in not a perceivable way, from the technological function after which every single building or every single area is used. To signal the function of every building through a new color code, could be the new way to read the system. The new city will have the task to bring back the intensity of the human presence to cover all city places. The city of the future must provoke, in passers by, the pleasure of the conquest of a goal and the emotion of observing. Who manages and will manage the cities, has and will have to find the courage to destroy, without mercy, those places that create sadness and to place those signs that will determine new distances and paths. The cities of the future will have to be covered, even in their simplicities, with color distinguishing one from the other. Perhaps also the architects will return to participate in the realization of these new areas that probably will be not called cities anymore.

AFFILIATION: Private architectural firm
COUNTRY:Italy/ 意大利

设计概念 "新千禧年都市"

建筑艺术的大动荡将是人与人之间强有力关系的开始。或许新的都市会遗忘过去的历史，就建筑而言，将建造科技住宅以延续新都市系统的发展。当科技越发达，越会成为都市发展的引导。虽然发展道路与新的行为因科技而更明显的显现其特性，但应避免人际关系的孤立。

都市的模式将越来越受控制以允许每个关系的增长，建构分隔场所的界限将召唤人们回来居住在新的开放城市，并且使科技的运用成为进行过程的辅助。新的都市将具备智慧型网络，容许与所有技术性灾难即时连接。复杂技术将延伸到对每件事的认知，构筑建筑将采用这样的构想。建筑语汇将被赋予多样性，并且不是以可感知的方式，而代之以每个建筑物或区域所采用的技术功能。

以新的颜色编码为每个建筑物标志机能后，将是阅读新系统的方法。新的都市被赋予新的人口密度。未来的都市必须激起对过客征服的快感与观察的情感。管理者将必须找到毫不留情地摧毁制造哀伤的地方的勇气，并且确立决定新远景与路途的目标。就算以最简单的形式，未来的都市也应该被彼此区别的颜色所覆盖。或许建筑师也将回归这些并不被称作都市的领域中并参与它的实现过程。[翻译　蔡咏岚]

Media-tion — A Digital Cinema Production Studio/
媒体化—数码电影制片工作室

DESIGNER: Jeffrey Morgan

The digital cinema production studio provides a place for the development and experimentation with a new genre of interactive film. Located in the unique urban fabric of Camden Town in northern London, the building mediates between the real and the virtual through the delineation of solid and void. The transparency of the vertical planes questions the traditional typology of film studio, opening up the interior and allowing visual access from the exterior. A void is created as the building pulls away from any direct contact with freight and passenger trains rumbling along the brick viaduct to the north of the site.

The urban context of the building plays an important role in the development of the design. Given the increased presence of digital communication technologies, the role of physical spaces and places becomes destabilized. We are no longer creating places of monumentality as in previous times. Rather, architecture functions more as a terminal for information, subject to changes in the field of information.

Given this, the building offers the possibility to engage in both physical space and virtual space. Finding the ways in which the physical and virtual may fuse together becomes the focus. The facades of the building provide the surfaces onto which the virtual spaces may be projected. By doing so, the limits of the facade are called into question, inviting further investigation into the design.

A system of delineation between the physical and the virtual was developed. The interwoven planes represent the solid, physical environment, while the vertical space of the facades create the space of the virtual. The intent was to create spaces that were separated not only by physical elements, but also digital representations and projections.

数码电影制片工作室是提供新型互动影片发展与实验所需空间的场所。基地位于伦敦北部 Camden 城镇，在独特的都市结构中，工作室凭着对实体、虚体的描绘调和了现实世界与虚拟世界。由于垂直面的穿透特性，释放了室内空间，并允许室外视线的介入，此举向传统制片工作室的空间模式提出了质疑。为回避北边高架道路上载货与载客车辆的噪音，脱开的建筑量体形成一个虚空间。

都市结构在数码电影制片工作室的设计发展上，扮演了重要的角色。由于数码传播技术的提高，真实空间的定位因而被撼动。我们不再创造纪念性空间了，更确切地说，建筑较像在资讯领域中随时变换的终端机。

因此，本建筑提供了同时投身于真实与虚拟空间的可能性，而找寻两者间融合的方法成为设计焦点。建筑的立面提供了虚拟空间可借以投影的面，这样的做法使得"立面界限是什么？"的问题浮上台面，引发对本设计做更深入的探讨。

真实与虚拟空间之间的轮廓描写系统逐渐成形。交织的平面表现实体的物理环境，而立面的垂直空间创造了虚拟的空间。我的意图是创造空间，这些空间不但是由真实元素加以区别，而且凭借数码元素象征和影像投映来呈现。[翻译 蔡咏岚]

↓ South Elevation from Regent's Canal／ 南向立面图

AFFILIATION: University of Illinois at Chicago
COUNTRY: U. S. A./ 美国

Main4mind. Homebase, A Future Virtual Environment for One Person/
个人的未来环境
DESIGNER: Kerstin Mueller

Cybertecture   The term "architecture" must be redefined in the context of virtual space, the search for a new definition leads to a new term - cybertecture. Some aspects of architecture are still valid in the virtual realm, like navigating a space. Other principles, such as the use of buildings to protect people from the elements, become meaningless. Completely new principles emerge due to things like the absence of gravity.

Project   The theoretical part of this work describes the spatial principles and possible shapes of virtual spaces according to their function. Insights from this examination are translated into a model of a future virtual environment for one person, the main4mind. homebase.

Main4mind. Homebase   [1] In the future, the virtual world and the physical world will have equal value. [2] Everyone will be the owner of a homebase. [3] A homebase is a virtual, personal environment for its user.

This project presents one of many possible homebases: main4mind. It contains four parts: [1] online-mode/ information access: The homebase can be used while active in the physical world. Virtual audio and images are superimposed over the actual world in a form of augmented reality. [2] wor-ld wide wor-k/ the world of work: This component serves as a working environment. To facilitate comprehension of the space, the environment is built in the form of a modular case system oriented on a simple cartesian grid. [3] my-artificial intelligence/ entertainment suited to the users' interests: Topics of interest are represented by spindles information units found by the main4mind software agent are positioned in-between the spindles, where proximity to a spindle indicates relevance to the spindle topic. The spindle positions are also modified to accomodate the changing quantity and distribution of information. Spindles provide a navigational reference for the user within the constantly changing structure. In this environment a user can both orient themselves, or get joyfully lost. [4] electronic-emotion/ an intimate place to meet: In a conversation, the surrounding physical space loses importance. The space created by body language and gesture between the interlocutors is important. These usually very fleeting actions become the space-building elements of e-motion.

网络建筑   在虚拟世界中，建筑被定义为网络建筑。传统建筑与网络建筑会有若干相似性，例如两者都游走于空间中。

网络建筑无须再像传统的建筑那样，需提供使用者的庇护所。重力在网络建筑中并不存在，也因此具备开发新准则的可能性。

此方案旨在探索虚拟空间的空间准则及其造型语汇，这些准则终会变成为 Main4mind. Homebase 个人量身订做的虚拟环境。

Main4mind. Homebase   [1] 在未来，虚拟世界的价值会等同真实世界。[2] 每个人均会拥有Home base。[3] 每个Home base 均为其使用者量身订做。

这些Home base 信息包括：
[1] 线上服务与信息的取得方式：虚拟世界的影音及图像在现实世界的应用已逐渐提高。[2] 工作空间：此元素提供一个工作环境，主要是正交时的格子系统，以便更了解空间。[3] 人工智能与娱乐系统：以条状图案代表不同主题，其间则是利用 main4mind 软件发现的信息，其位置并会依信息量而调整。使用者游走于空间时，可以用条状图案来定位。在这样的环境中使用者可自己定位，也可刻意享受迷路的乐趣。[4] 电子的情绪或私密的会面领域(聊天室)：谈话时，周围环境其实是不太重要的，重要的是谈话者身体语言所创造出的空间。这些姿势会在聊天空中(所谓的电子情绪)建构空间。[翻译 赵梦琳]

↑ My-artificial intelligence/ entertainment suited to the users' interests/ 使用者的兴趣

COUNTRY: Germany/ 德国

The Controlled Space/ 控制空间

DESIGNER: Gregor Hoheisel + Christoph Korner + Lars Krueckeberg + Wolfram Putz

The terms of "Public and Private", the traditional organizing principles of urban life, are out of focus, they are blurred into the realm of global networks, media outings of intimacy and new spaceless neighborhoods. Private goes public and vice versa.

Corporate interests are more and more controlling the public space with cameras, security services and surveillance. "Democratic Information" is unfree. Citizens are forced to decide to be either voyeurs or exhibitionists. The increasing violence in public spaces demands new solutions between controlling authority and personal freedom.

The project "The Controlled Space" explores the ambiguity of urban phenomenons in the public space.

Pub-lic adj. Abbr. pub [1] Of, concerning, or affecting the community of the people: the public good [2] Maintained for or used by the people or community: a public park [3] Participated in or attended by the people or community: public worship [4] Connected with or acting on behalf of the people, community or government, rather than private matters or interests: public office [5] Open to the knowledge or judgement of all; notorious: a public scandal.

Pri-vate adj. Abbr. priv., pvt. [1] Secluded from the sight, presence or intrusion of others: a private bathroom. [2]Of or confined to one person, personal: private opinions. [3] Not available for public use, control or participation: a private club [4] Belonging to a particular person or persons, as opposed to the public and government: private property [5] Not public, intimate: a secret

MADE surveyed the invisible space of surveillance cameras in four city blocks in Manhattan. The projects transformes the "observed volumes" of the inmaterial camera cones into visible projections onto the city's skin. In the zones of high involuntary public exhibitionism, the density of "docile space" is counterbalanced by the insertion of a public veil. This "urban skin" allows for a protected - "pagan" - space, where the nakedness of the urban traveller has been reversed and the innocence of public being has returned. Physical architectural manifestation becomes form by building an imprint of inmaterial digital space.

传统都市计划是以公私空间的分别为依据，此公私空间的界限已因网络媒体入侵私生活及无边界的社区而模糊。私人领域已公众化，反之亦然。

公共空间也逐渐被企业的摄像机与安全及监控系统所控制。信息更加民主化并不意味着人们更加自由。市民们不是得以偷窥便是被别人偷窥；公共空间内逐渐增加的暴力，迫使人们必须思考权威当局与个人自由之间的矛盾。

本控制空间即旨在探索这些公共空间暧昧的都市现象。

MADE工作室，在曼哈顿四个街区里架设了摄像机以监控看不见的空间。此设计者旨在将抽象的摄像机视野投射于城市表层。在这些强制性的公开展览中，空间会以一些公共性的屏障来平衡。这些城市表面包含了异教徒式的游乐空间，在这些游乐空间中，都市漫游者不再赤裸，城市的公共空间又恢复了原来的本色。建筑物其实会成为抽象的数码空间。[翻译赵梦琳]

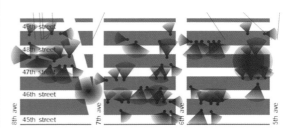

The Naked City Survey of the Controlled Space, Manhattan 2000/ 以控制空间来量度的赤裸城市

AFFILIATION: GRAFT with MADE
COUNTRY: U.S.A. + Germany / 美国 + 德国

UNE - UNE

DESIGNER: Masahiro Saigoh + Risa Kiyoshima + Atsushi Ezato + Mai Uchimura

The current library has been used only for the purpose to obtain information as looking for the book or studying. The library of the information age aims to be the stage of the exchange of the person and the person, the stage of the exchange of the person and information, and the stage of the exchange of information and information. The library produces new information by generating stimulation and touching off from the exchange. It is a base where the user not only obtains information but also processed information voluntarily, transmits information.

现在的图书馆只用于查询书籍获取信息或研读。信息时代的图书馆试图成为人与人、人与信息、信息与信息交流的舞台。图书馆通过产生刺激与触发交流而产生新信息。这里不只是使用者获得信息，也是他亲身处理信息与传输信息的场地。[翻译 蔡咏岚]

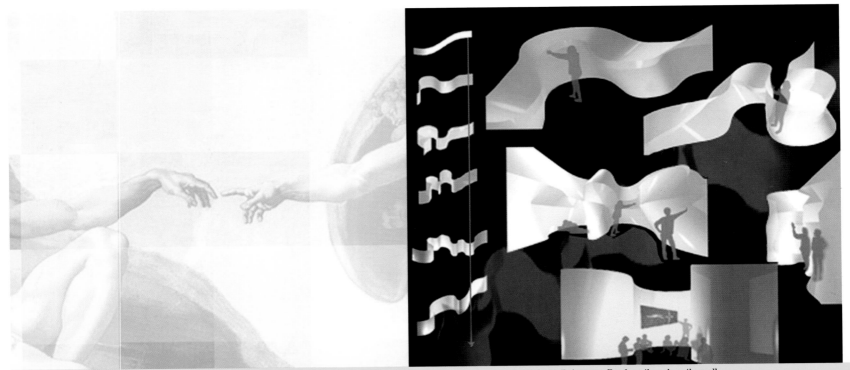

↑ Opening／ 开口　→The winding wall "UNE-UNE" is an axis of the exchange, passes the entire facilities, is moved by the visions, and forms their space, People gather along the wall, then, a new exchange will be generated.／ "UNE-UNE" 卷曲的墙是交流的轴、穿越整个设施，因影像而动，形成空间，渐渐卷曲。人们沿着墙聚集,然后产生新的交流

AFFILIATION: Assistant Prof. Soujyou Univ + Dept. of Architecture Graduate School of Science and Technology Kumamoto Univ.
COUNTRY: Japan/ 日本

Leisureland/ 城市乐园

DESIGNER: Andrew Vrana

The city, as we have come to understand it in the beginning of the 21st century, is a complex network of convergent vectors of material and information. In order to plan and design for the imminent changes that it will undergo in the next age, architects must harness the power of the new systems of organization that are emerging in the city and in cyberspace. The revolution of the electronic information age has paralleled the explosion of existing urban form. The city and its infrastructure must be rethought to cope with the current societal changes initiated by radical technological transformations. The microcosm of digital circuitry is causing profound effects on a global scale. Under these assumptions, TOPOS is a site of architectural experimentation that attempts to synthesize these issues of vastly differing scales. From the microprocesses of informatics to the global influence of interwoven metropolitan urban systems and vast virtual universe of the Internet, a new kind of architecture can be manifested.

The recent trend of rehabilitating waterfront real estate, in this case the reprogramming of an entire island abandoned by the military after the end of the Cold War in the middle of the New York metropolis, initiated the design of a complex of structures that are to serve the desires of the city for entertainment, gaming, and shopping while maintaining political distance from it. This "themed" landscape is more an intense receptacle of spectacles rather than a shallow veil of images and pure virtuality. Real-time events co-exist with the distraction of received and sampled information and experiences from the global network of entertainment/financial market speculation. The complex's proximity to Wall Street forms an intricate relationship between gaming (sporting events/gambling) and the Stock Market.

城市，一如我们在21世纪初所认知的，是物质与信息的聚合。为了掌握下个世纪城市的变化，建筑师必须能够控制城市组织及虚拟空间。电子信息带来的革命，会与现有的都市形态并存，因此城市与其公共设施必须配合由产业革命带来的快速变迁。数码电路带来的深刻影响是全球性的。TOPOS即为基于此种假设的建筑试验基地，会依不同的尺度来综合这些议题，这些尺度包括信息的微处理过程，一直到交融的都市系统在全球方面的影响，通过网络世界，将会有新的建筑产生。

近来在都市中，已逐渐开发水域的用途。此基地位于纽约市中心，是二战后废弃的军事基地。我们所设计的城市乐园，旨在满足市民休闲、游戏、购物等需求，但又与市中心保持一点距离。此乐园不仅是虚拟的意象，也是一处独特的景观。实际发生的事件，也正与由全球娱乐、财务、市场网络而来的各项信息并存，乐园与艺术街十分接近，更形成游戏与股票市场之间密切的关系。[翻译　赵梦琳]

↓ Mallscape/ 商场

AFFILIATION: TOPOS Architecture
COUNTRY: U. S. A. / 美国

Alternative Domestic Environment/ 另类居家环境

DESIGNER: Tony Youte

In the digital age, even though the visibility of electronic activity is minimal, comparing to the performance of walking in the dor of city's streets, the qualitative extension of privacy, is rooted in the unstable and unpredictable component of everyday life. Ambiguity becomes an attractive value, as it is able to federate contradictory, if not complementary desires: privacy/publicity. Open control operates the management of the situations.Hyper-functionalism is a neologism. The prefix Hyper is associated to the idea of functionalism to signify over-exaggeration applied to that concept. Hyper-here is more theatrical and deals with compositional talent and the achievement of effects by taking advantage of all contextual elements in a highly federative and integrative movement. Hyperfunctionalism offers the opportunity to merge the concepts of hard and soft technology with flexible sets of attributes, which address the convergent importance of aesthetic, behavioral, and comfort related concerns. The ideas that have generated the design of the Alternative Domestic Environment project were based on the theatricalization of density and on the relationship of the building to the sky. The presence of the building had to call the sense of contemplation, without unnecessarily monumentalizing the domestic character of the building and it's technology in the city's landscape. Beside the sophisticated arrangement of the programmatic elements (rooms, "regions of usage", circulation, facilities), I have created a federative element that poetically combines the connoted gravity of density with the lightness of the sky: The roofing-envelope. These elements acquire by their intricacy with highly poetically connoted features (the landscape, light, and sky), a second level of meaning. The mass-produced house offers optional modules and colors. The domestic environment is extensible and takes advantage of natural energy resources with the use of solar panels over its envelope.

在数码时代，虽然电子是无形的，却让我们的日常生活以及私密性得到改变。暧昧成为受欢迎的价值观，如同私密与公开之间的矛盾。超机能主义已成为新名词，所谓"超"字有点被过分强调：它其实是理论性的，利用涵构中的各种元素及其彼此间的互动，超机能主义可结合软硬件及弹性属性，强调美学行为学及其他相关议题的重要性。本设计旨在探讨建筑物与天空之间的关系，希望建筑物能唤起人们的思索，却又不致使城市景观具有过度纪念性。除了巧妙安排空间（房间、用途、动线、设备），我特别注意到物体的沉重与天空的轻巧之间的巧妙对话。或是第二层意义：像屋顶与景观、光线、天空之间的关系，房屋量体则呈现不同的模式与颜色。另类居家环境是可延伸的，并设有集能板以利用太阳能。[翻译 赵梦琳]

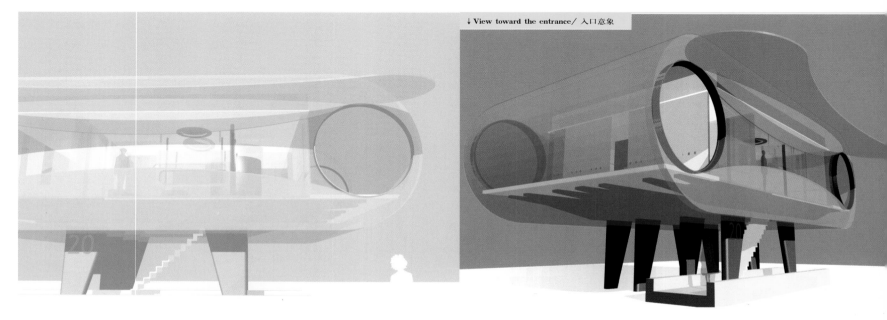

↓ View toward the entrance/ 入口意象

AFFILIATION: Architect/ Designer,Graduate Teaching Associate The Ohio State University
COUNTRY: Italy/ 意大利

Case Forces a Choice / 选择

DESIGNER: arch. Lisa Balasso

Design concept Which kind of architecture in the new era? In architecture we can say where are you going? Thinking of a real place! In the global network we know many virtual places. Always places for our feeling are both strong and real ... Architecture is changing as are the primary needs of our generation. Which is the gap between the real and the virtual places, if there is any? Where information is everything and the connectivity is primary.

Where the individual life became more and more important losing the real contact between souls.

Where everyone dreams of choosing, forgetting to be chosen by a cage of behavior and habits.

Where a button or a joystick is the means by which you can realize anything: our happiness and our destruction.

And we can say where is your body and where is your mind and both are far away and have a touchable feeling.

Places for the body have been built for ages but places for the mind have only been modeled, painted or played by an instrument now we also have to build these places exclusively for our mind. Looking at the reality that is loosing its substance. And thinking of a "fantasy" world which involves our existence.

Program Cords - Beginning for chance - Physical walls - doors - cages with many doors - in the "freedom" of choice - and then the eventual cords ... the sand.

新世纪将会有什么样的建筑？在建筑里你将去往何处？全球化的网络世界中有许多虚拟空间，但真实的空间又如何？会让我们有感觉的空间便是有力而真实的，当时代寻求改变，建筑也跟着改变，真实与虚拟场所的差别又在哪里？真的有差别吗？虚拟世界信息代表一切，沟通是最重要的。

个人越来越重要，彼此间的灵魂却不再沟通。每个人都急于选择，却忘记自己被行为及习惯选择。希望能找到一个按钮或魔杖，让我们能很快了解一切，了解我们的快乐与颓废。追寻你的身心，我们却深受感动，虽然身心已远离。

如果说我们一直建的是庇护身体的建筑，那么，庇护心灵的建筑只被油漆修缮而已。如果只看现实，现实就不会有意义，但勇于幻想，却能证明我们的存在。

空间计划：机会的开始—物质墙—许多门—选择的自由—最后的弦：沙地 [翻译 赵梦琳]

Image 1 Cords: where are you going? / 第一弦：你往何处去？

Image 2 Beginning for chance/ 机会的开始

Image 3 Phisical walls – doors –cages with many doors– where are you going?/ 实墙—门—为许多门所困—你往何处去？

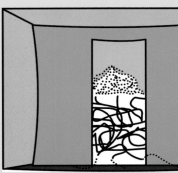

Image 6 and then the eventual cords ... t sand. / 最后的弦：沙地

AFFILIATION: None
COUNTRY: Italy/ 意大利

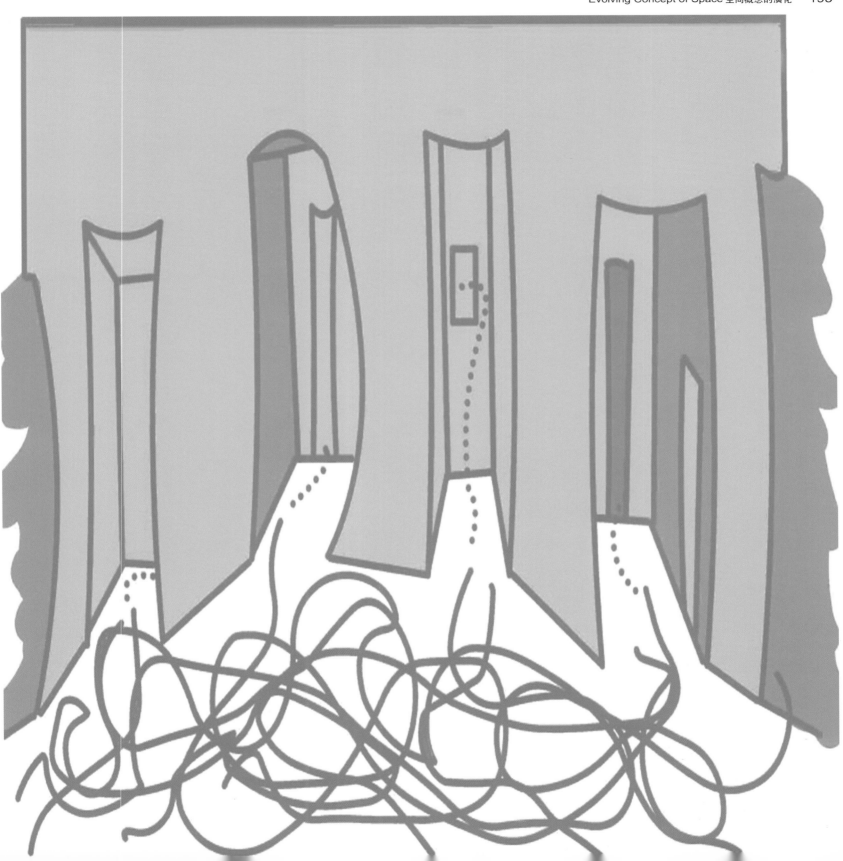

Representation of Digital Media

Representation of Digital Media   Linguistic research has showed that mathematicians purposefully or unwittingly use a dense language to describe simple mathematic concepts, leading to the commonly-held conception that only a few are able to master mathematics. Architects, for their part, use drawings and models to represent their designs. Most people are unable to read design plans and often are at a loss when looking at models. Thus, architects have continued to employ their "dense" materials for quite some time despite the fact that their designs are not that difficult to master. They have retained control over architectural knowledge, howeverl. With the development of digital architecture, in addition to the conceptual revolution it will bring, it will also provide ordinary people with tools for representing space. It remains to be seen, however, when the rapid growth of digital representing media will strengthen the monopoly of architectural knowledge by architects, or will instead open it up to the untrained.

数码媒体的呈现　　　语言学的研究曾指出,数学家有意无意地使用某些特定的艰涩的语法,来叙述实际上并不那么艰涩的

数学问题,因而形成"数学是少数人才能精通"的现象,以便长期把持知识权。因此数学家用某些特定媒体（艰涩语法）

来表现他们的想法,像建筑师用特定媒体（图画、模型等）来表现设计思想一样。由于一般人不容易读得懂设计图,看模型也经

常产生空间感的假象,因此建筑师是否也在长久以来,有意无意地使用"艰涩的媒体",来表现实际上并不那么艰涩的内容,以

便长期把持建筑的知识权,这个推测仍需社会与政治性的深入研究。但在数码建筑的发展上,除了思考媒体革命引起的形式与

空间的解放以外,这些数码媒体也提供了建筑师自己以及一般人能够十分清楚掌握空间的表现工具。这些数码表现媒体的快速发展,

到底会更强化建筑师长期垄断建筑知识的条件,还是有助于建筑师自己与一般人理解建筑设计,这些将值得我们在未来几年内

继续观察。

Living in the City / 城市生活

DESIGNER: Maarten Van Breman

A. Multipli-City   Trends in society can only be directed by bringing clear visions into practice through strategic corporate policies based on public support. The city of London is in need of projects taking new forms of urban life styles into consideration.

The project is explicit in its integration of different lifestyles and social groups within one Superblock. The organisational form and architectural aesthetics create a typical new urban fragment offering new forms of use and living in the city and opening new perspectives on social and economic sustainability and cohesion. Environmental sustainability will be reached by use of innovative materials and installation technology, building methods.

Trends — A series of powerful trends can be traced that influence our way of living in the city. The urban environment forms the battle-field for this New Economy under construction and will change rapidly along with it.

Themes — Projects for urban development should anticipate on these revolutionary changes in society. Our urban environment, challenged by the 'Zeitgeist', should offer alternatives by providing conditions for new developments in contemporary life-styles. Relevant themes related to our profession should be defined.

Lifestyles — Research on our urban population is principally based on a breakdown in social groups by household types and financial income. More accurate categories related to urban development would be diversification in lifestyles showing characteristics in social behavior, demand for specific living space and urban program.

[1] From Household to Lifestyle: Projected breakdown of households, Calculation of households in m2 on site, Projected breakdown in lifestyles on site [2] 24 Hour Lifecycle of the Block: Inhabitants, Cave behavior, Synthesis: Multipli-city

B. City within the City   Basically the site at Bishopsgate Goodsyard of 2.0 hectares will be occupied by one dense Superblock placed on the existing socle.The sculptural block forms a characteristic new urban fragment offering a variety of housing typologies (62.000 m2 living space for about 1400 people) and additional urban program. Its central interior area 'the city cave' creates a new urban world in addition to the city itself and the planned open landscape park next to it. Its urban program and inhabitants ensure a 24 hour lively part of the city fabric. The more the block appears as a city within the city.

C. Design Principles   The architectural form and organisation of the block is created by a series of transformations and design principles. The general themes observed as relevant to urban change are present in these principles, culminating in a complex multipli-city.

Densification — "the layered city"   The Superblock: a dense urban fragment, "Stuffing" the city cave with "dark" urban programme.

Synergy — "the city as a melting pot"   "Slicing & shaping" the block into of a series of slabs, "Labelling" the slabs with lifestyles and typologies, Synergy of the diverse interrelated program.

A.复合式城市   社会的整体趋势，由公共政策造成，公共政策必须有明晰的远见及民意基础。目前伦敦所需的建设，必须建构于新的城市生活之上。

此方案位于一超大街区上，结合不同的社会群众及其不同的生活方式。设计者以一种新的空间组织与建筑美学观，创造具有代表性的新都市生活，提供市民以新的都市机能、社会凝聚力及都市生活。至于可持续性发展，则有赖于新的建材及建筑技术。

趋势—影响城市生活有好几个趋势。城市的环境，基本上是新经济的战场，并会随着新经济而改变。

主题—都市发展反映社会革命性的变迁，我们的都市环境应受时代精神与当代生活方式的挑战，并重新定义出我们生活的主题。

生活方式—我们是以生活方式及收入状况来区分社会人口的，但进一步的分类需依据社会行为，对空间的要求和都市计划。

[1] 从家族到生活方式:不同家庭形态的分类、不同住宅所占的建筑面积、城市内部不同的生活方式。[2] 城市内夜以继日的活动：居民、城市中心区的居住方式、合成构成：复合式的城市。

B.城中城   城市位于 Bishopgate Goodsyard ，占地两公顷，被设定为惟一的一

AFFILIATION: Group for Architecture
COUNTRY: Netherlands /荷兰

Networking — "the networked city"   Getting the internal infrastructure connected with the urban context. The urban program in the block serving a greater urban area.

Sustainability — "the eternal city"   Ensuring social sustainability within the block and the public realm. Specification of the voids: ecological themes and urban program. 24 hour activities and economic sustainability of the Superblock.

New Policies — "the managed urban development"   New policies in public-private development and investment. Ownership per unit, per slab and tenure for the whole block. [Text provided by Group A]

个超大街区。此超大街廊包含各类不同的住宅形态（建筑面积 62000 平方米，居住人口 1400人）及其他的空间。街区中有一相比邻的公园，即所谓城中城的区域，可供给市民24小时健康的组织生活，更符合城中城的性质。

C.设计原则   建筑形状及街区的空间结构是由一系列变化的设计原则组成，这些原则主要基于城市的复合性。

密集—"多层次的城市"   超大街区：高密度的都市集合体，以夜生活的生活模式来填充市中心。

↑Longitudinal Elevation ／ 长向立面 ↓Site Plan ／ 全区配置

合成—"城市如熔炉" 将街区重组为一系列的楼层，将楼层分类为不同的生活方式与类型、不同却又相互关联的建筑计划的合成。

连接—"连接不同的城市" 将城市与都市结构结合，都市计划就可照顾更大的都会区。

持续性—"永恒的城市" 确定街区及公共领域的持续性。刻意定义的虚空间：以生态与都市计划为主，超大街区24小时的经济活动具有持续性。

新政策—"有计划的城市开发" 都市公私部门的投资发展政策和每单元楼层的产权。

[翻译 赵梦琳]

←Climbers Garden Slab ／ 登山者的花园区 ↓Interior Space ／ 内部空间

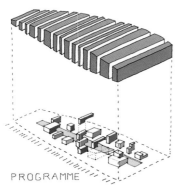

"密集"分析图／ Construing illustration of "Densification"

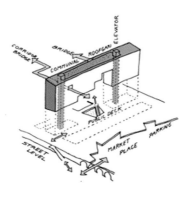

"连接"分析图／ Construing illustration of "Networking"

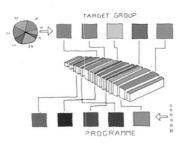

"合成"分析图／ Construing illustration of "Synergy"

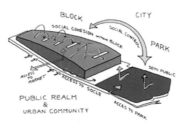

"持续性"分析图／ Construing illustration of "Sustainability"

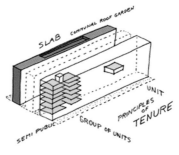

"新政策"分析图／ Construing illustration of "New Policies"

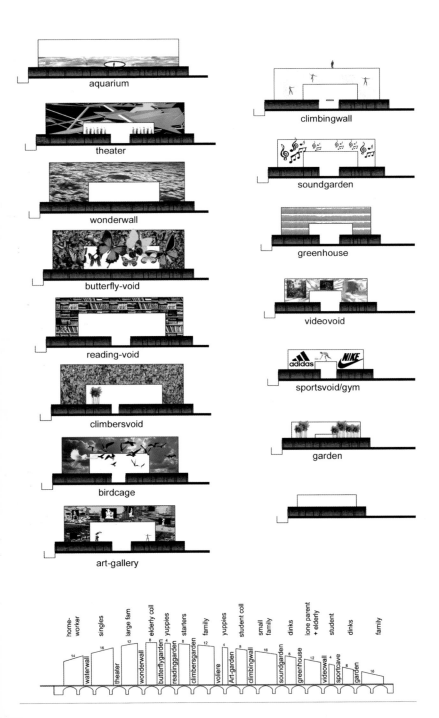

aquarium

theater

wonderwall

butterfly-void

reading-void

climbersvoid

birdcage

art-gallery

climbingwall

soundgarden

greenhouse

videovoid

sportsvoid/gym

garden

←By moving through series of slabs one can experience a cross section of our contemporary multi-cultured society 穿越一系列的垂直片段，在横向剖面体验当代多元文化社会

Borderline — Complexity and Digital Media / 边界—复杂性与数码媒体
DESIGNERS: Laura Cantarella

One of the major topics about contemporary city's territorial transfomations is the question of uncertainty. Constant shifts in territorial, economical and geopolitical identity coexist with today's incapacity and lack of tools in observing and representing the dynamics that are actually redrawing political and economical geography. Stratifications and destratifications, nonlinear multiple interactions of the many individuals and structures, many and divergent evolutive forces create what we could call accumulated complexity. The contemporary city does not follow in its evolution a linear movement, where the succession of elements is distinct and the causes clearly identifiable in their nature.

The transformations that mark it are directed towards distant and concurrent goals, promoted by a multitude of actors which interact without knowledge of the overall situation. For this plural nature, the contemporary city has a dynamic and ever-changing behaviour, seat of an infinity of different evolutive rhythms, of gradual transformations, of slow shifts in the cognitive maps of its inhabitants, of periodical amplifications of activities in some environments, of instantaneous retractions from inhabited areas, of slow implementation of some infrastructural systems, of never-ceasing redefinition of combinatorial criteria of public and private, of a cyclical alternance of intensity of use of the space of public life, of a syncopated shifting of the shared spaces. The duality global/local is not simply an opposition, but expresses the component parts of a single process of stabilizing the new modernity.

Criteria   Borderline - Complexity and Digital Media is not a project about a city. It is not a project at all. It's a reflection, a suggestion about the city at this borderline time: the policentric city, the multiethnic city, the dence city, the communication city, the hypertext city, the democratic city, the antimonumental city, the virtual city, the information city, the surface city, the media city, the interactive city, the nomadic city, the transition city, the shelter city, the political city, the people city, the borderline city. The question is: how can digital media say something about this contemporary complexity? This investigation was made by the use of collage. The most famous example of collage as a form of expression is probably Picasso's cubist painting. In architecture history an interesting sperimentation was made by Archigram in the 60s and nowadays by contemporary architect Rem Koolhas. Another source of inspiration is Marcel Duchamp's "ready made" concept. Any object, moved from his original position, gets another meaning. Moreover putting togheter different images everyone can recreate meanings, saying something completely different from the original message for which they were made.

The collage is a way to say that we cannot explain anything with sharp single images anymore, but multiple, blurred, differnt ones. cutting images is a way of understanding that is the first step in order to take care of something. We may move in contemporary cities like space detecitves, collecting images and words, trying to say something putting together what we collect. Images coexist and interact with text, they are on the same level. This digital space is made up of endlessly proliferating meanings which have no stable point of origin or closure. The boundaries which enclose the work are dissolved. The text opens continuously into other texts and other images - the space of intertextuality.

Quotation   "Looking at the boundless world that explodes within territories, we must look for a new common language, aware that every single word, sound, figure, photography, painting, have become entire costellations of meanings and references, echo and mixtures of different forms and beings"   — L. Ghirri

现代都市领域性变化的主题之一是关于不确定性的疑问。领域性、经济性和地理性持续的认同与变化，与今天对于观察重塑政治和经济形势的动态工具的缺乏并存。阶层的形成与瓦解，复数个体与构造的非线性交互影响，许多相异渐进的力量创造了我们所称的累积复杂性。现代都市并非沿着线性发展，在发展的运动中元素的接续是独特的并且其结果可清楚地辨识出来。

其变化指向遥远一致的目标，这是由许多不完全了解情况的参与者促使而成。对此复数的特性，现代都市呈现动态持续变化的特点，容纳永恒多变的演化节奏，缓慢的转化，居民认知地图的迟缓变换，某些环境中活动的阶段性扩大，居住区域的瞬间收缩，一些公共设施系统缓慢的完成，公共与私人间永不停歇组合准则的反复定义。公共生活空间环形交互作用的强度，共有空间切分的转变。全球与区域的二元性不仅是敌对的，也传达了稳固新现代性单一过程的组成。

评断准则   "边界—复杂性与数码媒体"并非关于都市的提案。它甚至不是个提案而是个反映，是对于正处于边界的都市的建议：多民族都市、藏身的都市、沟通的都市、虚拟的都市、信息的都市、表面的都市、媒体的都市、互动的都市、流浪的都市、过度的都市、庇护所的都市、政治的都市、人民的都市、边界的都市。问题在于：数码媒体要如何对当代的复杂性提出意见？此次的探索以拼贴

AFFILIATION: Facolta' di architettura
COUNTRY: Italy / 意大利

的方式进行。最具代表性的拼贴作品应属毕加索的立体派绘画。在建筑历史中，20世纪60年代有关于建筑图像的有趣的实验，在现在则是当代建筑师Rem Koolhas所完成的。另一个启发的是马塞尔·杜尚的现成(ready made)的概念。任何自原位置移动的物体，会获得另一个涵义。组合不同影像，大家可以重创涵义，表达完全不同的原始意念。

拼贴是说明我们不能继续以单一的而应以多重、模糊的、相异的影像来解释任何事物的方法，裁剪影像以了解它是处理事情的第一步方式。我们或许会像空间侦探一样在现代都市中移动，搜集影像与字句，试图组合所搜集的

信息表达了些什么。影像与文字互相依存并相互作用，它们处于同等地位。此一数码空间由没有稳定原点或封闭无尽激增的涵义组合而成。围绕作品的界限消失了。文字不停地与其他文字和其他影像形成了一个融合交织的空间。

引用 "身处于有限的时空中，我们的目光却投向那无尽的世界；我们急需一种新的通用的语言，它能让每一个字、每一个发音、每一个图形、每一幅照片、每一幅画以及每一个回声和所有的存在，都成为夜空中涵义深邃的不落星辰。" — L. Ghirri   [翻译 蔡咏岚]

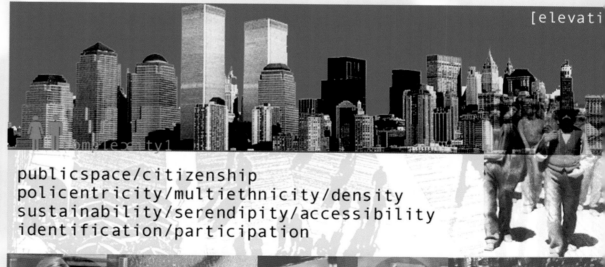

publicspace/citizenship
policentricity/multiethnicity/density
sustainability/serendipity/accessibility
identification/participation

→Communication, democracy, antimonumental, establishment /
沟通、民主、非纪念性、建立

complexcity2

interface/surface/membrane
information/communication

[communication city]
[project space]
[city like hypertext]
[democracy and perception]
[public space]
[renewal space]
[establishment and antiestablishment]
[partecipation and imposition]
[antimonumental or monumental]

→ Borderline, nomadic space, experimentation,
indetermination / 分界线、流浪空间、实验、不确定性

flexibilit
temporar
ethic
shelte
politic

[borderlin
[space of indeterminatio
[landscape of transitio
[potencial of experimentatio
[absence of phisical and mental border
[places where uncertainties interac
[seismographs of transformatio
[theater of new and old nomadism
[continuity made up with fragment
[space of nomadic architectur

→Public space, policentricity, serendipity,
partecipation / 公共空间、意外收获、参与

Play Zone, Millennium Dome / 游戏区,千禧穹顶
DESIGNERS: Land Design Studio (Design), Robin Clark (Project Designer), Anthony Pearson (Computer Modeling)

Designing the Play zone at the Millennium Dome London, gave us the unique opportunity to employ computer modeling to design and develop a purpose built structure dedicated to the work of worldwide new media / digital artists. It is fascinating that this unprecedented showcase of an emerging digitally interactive genre has been enabled by using the constituent technology in a different format. We employed 3D computer modeling as an integral design tool at every stage of the project. As interpretive architects, we always endeavor to inform the external envelope from the internal narrative function of our installations. In this case a very early organising principle, was the desire to accommodate cantilevered back projection facilities for the extensive amount of data projection equipment that was demanded by the curated work.

The primary conical component was carefully designed and modeled in 3D to constantly evaluate and predict the ultimate finished product. This was especially important as this pre-fabricated repeat module created the distinctive external sculptural form of the structure. The 3D digital model also enabled easy reconfiguration of the complex internal sequencing and movement system, that were compliant with the manipulation and development of technical and aesthetic aspects of the projection cones. The speed at which the design tool could be used supported the need for us to be able to design, install and commission this complete 10 million entertainment venue within 350 days.

High quality rendering of the finished product was a major requirement to demonstrate concepts to an inexperienced client team and information hungry media. It is ironic that once constructed, the "digital playground" provided a sophisticated mechanism to demonstrate the work of digital artists who understand conceptual and cerebral ideas together with wit and humour, heavily underpinned by considerable computer skills.

One overwhelming objective was to introduce an interactive experience that would be intriguing, charming and simple to engage with. 18 games were installed that fulfilled the need to demonstrate play activity in terms of physical challenge, encouragement of self-expression and the cognitive ability to discover. Contrary to the adverse publicity the Millennium Dome has been subjected to, the Play zone has been voted best zone by opinion polls, acknowledged by D&AD and awarded "Best Entertainment Venue" by FX magazine.

AFFILIATION: Land Design Studio Ltd.
COUNTRY: U.K. / 英国

设计伦敦千禧穹顶的游戏区，让我们有机会将电脑模拟运用在献给世界新媒体—数码艺术家的设计与建构上。这个关于新兴数码互动艺术品的陈列是由基本科技以不同方式使用而成。我们运用 3D 电脑模拟作为每一阶段不可缺少的设计工具。身为诠释人的建筑师，我们总是力图由内在机能发展外观。在这种情况下，较早的组织准则是试图容纳大规模数据装备的悬臂后部署计划设施。

一个关键的圆锥形构件被小心地设计出来并进行了 3D 模拟，以期不断衡量与评估最终成果。因为此预先组装的重复模式所创造出构造体外部独特的雕塑形体特别重要。 3D 模型也让复合体内部序列与运动系统可以轻易地重新装配，它们顺应了圆锥体的技术与美学方面的处理与发展。此设计工具的操作速度让我们在 350 天内就完成了这 100 万娱乐集合场的设计安装与发包。

完成高品质的彩现是将概念呈现给一个没有经验的业主团队和媒体的重要条件。具有讽刺意味的是建造的"游戏区"，示范了机智风趣与想法理智的数位艺术家的作品，其作品的结构运用了大量电脑技术。

其中一个最关键的目标是介绍一个富于启发性的、迷人的、简洁的互动经验。安装了可以满足体能挑战、鼓励自我表达与探索能力的18个游戏。与千禧穹顶不同的是，游戏区被选为最受欢迎的区，被授予由 D&AD 公认和 FX 杂志颁发的"最佳娱乐场所"。[翻译 蔡咏岚]

Computer generated view of building entrance with "Big Neon Arrow" / 电脑模拟有"大霓虹箭头"的建筑入口

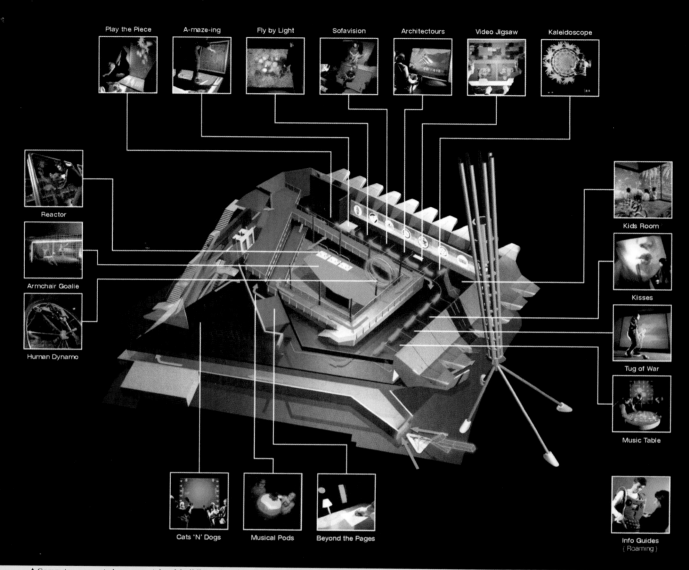

Play the Piece

A-maze-ing

Fly by Light

Sofavision

Architectours

Video Jigsaw

Kaleidoscope

Reactor

Armchair Goalie

Human Dynamo

Kids Room

Kisses

Tug of War

Music Table

Cats 'N' Dogs

Musical Pods

Beyond the Pages

Info Guides
(Roaming)

↑ Computer generated axonometric of building, showing layout of the 18 new media interactive games ／ 电脑模拟的建筑外部透视,显示18个互动新媒体游戏的配置
↓ Photographs of interior with games installed ／ 安装了游戏的室内照片

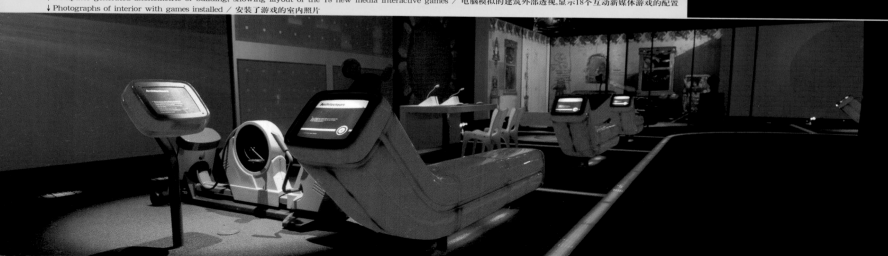

H-city — Urban Landscapes of Communication and Exchange
H 市—沟通与交流的都市地景

DESIGNERS & AFFILIATION: Boris Brorman (Architect MAA), Claus Peder Pedersen (Architect MAA, Ph. D.),
Claudia Antonia Morales (Carbone Architect MAA), Morten Daugaard (Cand. Phil.), Peter Hemmersam (Architect MAA),
Tom Nielsen (Architect MAA), Lars F.G. Bendrup (Architect MAA), Thomas Larsen (Architect MAA)

The H-city is an urban project on a national scale. The main framework of the project is connected to the concept of the H-city. H-city has been in the making since the sixties, with the planning of the motorway system parallel to the main railwaylines. It has been an issue in all general development schemes in Denmark ever since. Often with a focus on how it to deal with the distorted map of Denmark that pops up as a consequence. However the real potential in the concept became clear to everybody in 1998, with the opening of the Great Belt-Bridge. This 18 km.s of infrastructure more than doubled the amount of traffic travelling from East to West-Denmark, from 2.8 to 6.5 million cars. The "H" includes the 5 biggest cities in Denmark.

Around the "H" lives 2/3s of the countrys population and the relative proportion is still growing. Now with the new bridge to Sweden (2000) the H-city is closely connected to what is fastly becoming the second biggest city in Sweden, Malmo. The absurdity in the continuing "good intentions" about taking measures preventing the further "distortion of the map" is becoming still more evident.

Since the sixties the relative size of population, income, commercial and cultural developement has been growing in the "H" compared to the rest of the country in spite of all the measures taken from the Welfare State trying to contain this developement. The European perspective is most likely to intensify this overall bias; tendencies it is increasingly necessary to find a way to come to terms with. This project starts from a pragmatic acceptance of the built environment and the massive changes initiated by major infrastructural projects in Denmark during the recent years (primarily the new bridges). The project mixes analytic work with "imagineering" and actual architectonic proposals. This crosscultural and mixed media approach relies heavily on digital tools. Technologies like Photoshop and Illustrator allows for sampling,

superimposition and scaling of the percieved reality to new not yet realised realities. These tools, rather than Maya and AutoCAD, admit for an architecture formulated as an active interplay between the built and the cultural landscape. An interplay that underscores the continuing existence of the well known images of the danish landscape with minor adjustments of the built : the new windmills, high voltage lines, railroads and highways, new bridgemonuments together with old churchtowers. This is possible through an architecture that doesn't start with obvious problems, but tries to take advantage of the existing cultural and architectural values in relation to that cultural landscape we have chosen to label nature. An architecture that recognizes that construction of images of reality is reality and reality is construction.

Infrastructure is the point of departure of the H-city, a city of movement and time rather than space. This implies building along the main arteries of infrastructure instead of building and repairing on an old urban concept inherited in the latest version from the city of industrialism, hereby focusing on collaboration between different parts of the city and a better exploitation of ressources. The future urban diversity though as a whole: all urban areas should not necessarily have a little of everything but something substantial of different things. H-city bases its existence on this fact, not as a problem, but as a potential.

Its specific quality lies in the programmatic variation it is capable of sustaining, and on its introduction of landscape as an element within the city, restating the old fact that landscape and the city are two sides of the same coin. It is therefore

H 市是国际尺度的都市计划。该项目的框架与此都市的概念相联系。H 市 20 世纪 60 年代就开始建设，并把高速公路与主要铁路并行规划。丹麦一直以来所有发展方案的问题，主要集中于讨论如何处理丹麦歪曲的地图。但此概念真正的潜力是在1998年因为Great Belt桥的启用而变得明朗起来。这18公里的设施把东西丹麦间的交通量增加了两倍多，从 28 万到 65 万车次。H 市包括了丹麦5个最大的城市。环绕H市入住了丹麦三分之二的人口，而且还在增加。现在有通往瑞典的新桥，将H市与正在成为瑞典第二大城的Malmo市紧紧的连接起来。

防止 "版图的歪斜" 的加剧所采取善意的荒诞手段越来越明显。从 20 世纪 60 年代以来，H 市人口收入、商业与文化发展的规模都大于丹麦的其他区域，而不管福利部门为维持此发展所采取的措施。欧洲在未来很有可能加剧此这种歪斜。

此方案由计划性地接受已建环境与主要设施大规模转变开始(主要是新建的桥梁)。它揉合了分析想像和真实的建筑提案。此跨文化与揉合的媒体方法依赖于数码工具。像是Photoshop或Illustrator的技术让测试、附加与度量未发表的现实成为可行。这些工具容许建筑系统化地阐述与文化景观之间动态的互动。一个熟悉的丹麦景观在只做了微小变动的情况下强调其历史的延续性：新的风车、高压电缆、铁路与高速公路，新的纪念物与旧的教堂高塔。建筑试着利用现存的文化与建筑价值和

able to incorporate the many unpredictable forces and mechanisms in contemporary urban development, and has the ability to deal with the many unknown futures. It is the city as a "loose fit", rather than a fixed form.

H-city remembers the fact that all cities are based on the principle of accessibility, not as proximity and space as in the historical town or city, but as movement and information. This project on the H-city opens up for the fact that differentiated measures has to be taken if new development should be possible as in:

At the coutryside LITE(TM) At the coutryside LITE(TM) shows new possibilities for housing in postindustrial landscapes lied fallow by presenting a combination of preindustrial and advanced technological solutions. An economically and ecologically profitable framework based on gravelroads, ecological water supply, celluar phones and "on demand" buses around detached modern housing-units with possibilities for modest amounts of specialized farming.

At the countryside LITE(TM) distinguishes itself from At the countryside CLASSIC(TM) Through a deeper connectedness towards the surrounding world and a less sentimental relation to surrounding landscape characterized by planting of forest, landscapes of energy, landscapes of leisure etc.

文化景观的关系而成为可能，认识到建构现实的景象就是真实。

H市是一个动态运动于时间而非空间的都市。这意味着沿设施的主要干道建构而并不是在旧的都市基础上加以完善，在此着重都市不同部分的合作以及资源更好的开发上。未来都市的多样性虽是整体的但所有都市区域不必面面俱到，而要求各有特色。H市的生存基于这个原则，不视为问题，而视之为潜力。

这一特质存在于它所能维持的纲要变异之中，以及将景观引荐为都市元素的一部份，重新声明景观与都市是"手心与手背的关系"这一古老的观点。所以它可以混合现代都市规划中许多不可预知的力量与机制，并且有处理许多未知未来的能力。它是宽松而非固定形式的。

H市乘承了所有都市都基于可即性原则的事实，而非传统城镇或都市的邻近性或空间本身，是关于活动与信息的。本方案若有新发展，就需要接受不标准的事实：

在乡间 LITE（TM） 通过前工业时期与先进科技的混合，显示了在后工业地景新的居住可能。经济上和生态上有力的架构是生态的水源，移动电话与"招手停"公共汽车和环绕分离的现代住屋单元中，其中可能有小量的专业农作。

在乡间 LITE（TM）与乡间 CLASSIC（TM） 通过对于环境更深层次的连接以及与环绕地景较理性的关系而有所区别，以林木种植、能源景观、休闲景观等等为特征。[翻译 蔡咏岚]

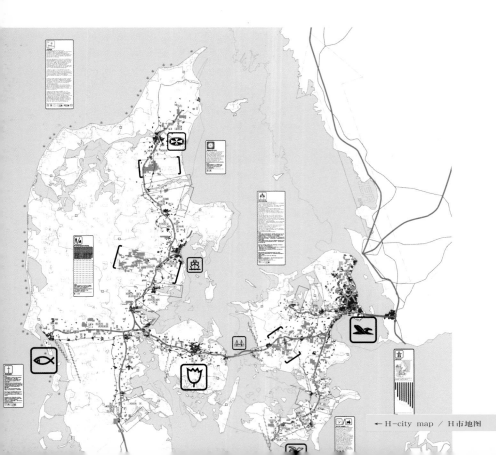

← H-city map / H市地图

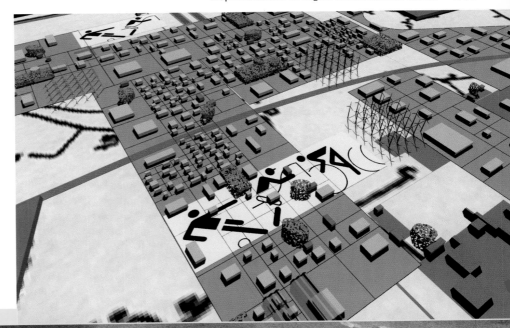

←Great Belt-Bridge / Great Belt桥　→Landscapes of leisure / 休闲的景观

↓ ↑ Imagenary constructed landscape / 想像的地景

National Public Library / 国家图书馆

DESIGNER & AFFILIATION: Carl R. Tully(NBBJ), Dace A. Campbell(NBBJ and the UW Human Interface Technology Lab),
Mark Farrelly(UWIRED and the UW Human Interface Technology Lab), Susan Campbell(Zaaz and the UW Human Interface Technology Lab),
Bruce Campbell(The UW Human Interface Technology Lab), Taylor Simpson(NBBJ)

INTRODUCTION　　In the early 21st century, all media will be distributed digitally, and stored and viewed using electronic books and holographic displays. The National Public Library (NPL) will provide free remote and local access to digital media. This NPL will be a "cybrid" institution, with virtual and physical components:

Virtual Library: accessible from any networked computer to provide a spatial interface to the digital collection by organizing digital information in 3D.

Information Kiosks: distributed throughout the public realm to provide convenient access to the NPL system.

Physical Branch Libraries: located in every city to give local patrons access to digital media and to provide a community center.

Furthermore, social program elements of the NPL exist both in physical branch libraries and on-line, creating a local community center for each city with unique ties to a networked virtual community.

Virtual Library　　The Virtual Library provides easy, intuitive access to a 3-D search space of the digital collection. An architectural, spatial metaphor allows patrons to visualize information and "meet" in social areas. Program items include the following, many of which have matching counterparts in the physical branch libraries:

Lobby / Information area / Chat space / Lecture auditorium / Search space / Meeting spaces

Information Kiosk　　The NPL Information Kiosks enable free, 24-hour access to the digital collection and are found in the reference areas of NPL branch libraries, as well as major public facilities, and public sidewalks. Each kiosk features advanced 3-D interface technology, configurable to any user's needs. Patrons insert their DataCards to login to the Virtual Library, and are immediately connected to the NPL network with the display automatically configured to their preferences.

Physical Library　　The NPL branch libraries offer state-of-the-art viewing interfaces in a comfortable, friendly space. Branches offer local patrons convenient access to the digital collection and a place for both community events and quiet, personal research. Program items include the following, many of which have matching counterparts in the virtual library:

Lobby / Information and circulation / Cafe / Lecture auditorium / Meeting rooms / Reference area / Childrens area / Reading area / Administration offices / Services

COUNTRY: U.S.A. / 美国

简介　　21 世纪初，所有媒体将会用电子书与全息摄影显示电脑系统化的分类、储存与阅览。国家图书馆将提供对数码媒体的免费遥控与地方性使用。国家图书馆将是电子化的机构，拥有虚拟设备与实体设备：

虚拟图书馆：从任何电脑网络皆可进入，以 3D 组织数码信息，提供数码馆藏的空间界面。

信息站：分布在公共范围内提供对于国家图书馆系统站的便利使用。

实体图书分馆：位于每个都市中，提供给地方人士一个数码媒体的使用并进行交流的中心。

此外，国家图书馆计划的社会性元素存在于实体图书分馆中与网络上，创造每个都市中紧系着虚拟社区的地方社区中心。

虚拟图书馆　　虚拟图书馆提供对于数码馆藏的 3D 搜寻空间的便利与人性化的使用。一个建筑的空间隐喻让参与者将其视觉化并在社交场合相遇。计划项目包括，其中有许多与实体图书分馆互相对应的部分：大厅、会议区、聊天室、演讲厅、搜寻场所、会议场所。

信息站　　位于参考区提供 24 小时免费信息，也有主要的公共设施与公共路。每个信息站都有 3D 界面科技的特点，依使用者需要可编配。参与者插入他们的资料卡以登录虚拟图书馆，并即刻连上图书馆然后依照他们各自的喜好自动编配信息网络画面。

View of an information search within the 3-D search space / 在搜寻区里的信息搜寻区视景

实体图书馆 国家图书分馆在一个舒适的空间中提供最先进的观赏界面。分馆提供给地方人士便利的馆藏使用、社区活动和安静的个人研究空间。计划项目包括大厅、信息与流通、咖啡厅、演讲厅、会议厅、参考区、儿童区、阅览区、行政办公室、服务区，其中许多与虚拟图书馆有相对应的部分。 [翻译 蔡咏岚]

The NPL information Kiosk / 国家图书馆信息站

← View of the kiosk in use within the reference area of a NPL physical branch library ／ 在国家图书分馆参考区使用中的信息站视景

↑ The reading area of a physical branch library ／ 实体图书分馆的阅览区

↓ A physical branch of the NPL as seen from the street ／ 国家图书馆的实体分馆从街上的视景

Project of Urban Rehabilitation in les Abymes Guadeloupe
位于 Guadeloupe 市 les Abymes 区的城市住宅更新

DESIGNER: Amgod#8, Alessandro Baldoni, Sabrina Cantalini, Giuseppe Catania, Stefania Ciuffoli, Adelaide Di Michele, Adriana Feo, Gabriele Mastrigli, Giuseppe Troccoli

The district of les Abymes in Guadeloupe is a contemporary part of town formed by an uncotrolled mix of different kinds of settlements, functions, infrastructures, voids, produced by mostly autodeterminated mechanisms of urban transformation. The heterogeneity of role, function, scale, perceptibility generated by this complexity of factors is assumed by the project as an element of quality, especially in the forms it takes in the selfbuilt zones, particularry meaningful for their system of internal and external relationship and for the richness and flexibility of their fragmented morphology.

So the project individuates a sequence of transverse bands with fragmented and irregular boundaries; they do not have in the road a place of identity; they do not have "fronts" or, at least, "street fronts": their relationship with the main roads is casual, strictly functional. Framed by the vehicular infrastructures, they are kept toghether from a continuous, interstitial void that constitutes the connecting space.

Trying to frame this void mantaining at the same time the maximum degree of freedom and flexibility of the settling process, the project proposes to build mainly the frame of the dwelling system consisting in the technological networks (energy, water, sewerage...) and in the network of public facilities; on this frame will set out, later on, the construction of the buildings with times and modes the more flexible as possible. So on the technological spine are built the main facilities (kitchen and bathrooms) with a steady structure (reinforced concrete or local bricks). Starting from them the inhabitants can self-build their houses with scraps or recycled materials, drawing independently their boundaries, to follow up the image of a protected place, strongly characterized at an individual level and rather introverted, creating a various, anarchist landscape, changing frequently, inside a never-ending process of destroying and reconstructing the space that mantains to this site his strong carachter of vitality.

Guadeloupe 市的 les Abymes 区由近代未经规划的各种建筑物、功能机构、公共建设和虚空间形成，大都是由自我界定的都市转型的机械主义制造的产物。

在角色、尺度和可辨性上的异质性被视为本案的特点，尤其在已有建筑的区域中的造型，对于它们内在与外在关系以及破碎形态的丰富与弹性是尤其有意思的。所以本案将一连串横向圈带与破碎不规则的界线个性化。它们的街道没有个性，没有正面，甚至没有街面；与主要街道的关系是随便的，纯粹机能性的。它们被交通设施框住而合起来形成连接空间的连续有缝隙的虚拟空间。

构成此虚拟空间的同时需要维持开发最大的自由度与弹性，本方案提议主要建造的住宅系统包含技术性网络（能源、供水、排污…）和公共设施；稍后在此框架上将建造尽可能有弹性的建筑。技术性的文体框架继续建造，包含有稳定构造（加固混凝土或本地的砖）的厨房和浴室。从这个框架的基础上空间住户可以自己用废料或再生材料规划他们的家，形成私人空间，通过强烈将个体的特性描绘出来而非内向含蓄的，制造一个多样化的自由景观，通过经常变换，在内部形成不停歇的破坏与重建充满活力的空间。 ［翻译 蔡咏岚］

↑ The networks of facilities and public spaces framing the private spaces ／ 框着私人空间的公共设施与公共空间的网络

↑ The semi public among the complex system of the self-built dwellings ／ 在自建住宅复杂系统中的半公共空间

AFFILIATION: Amgod#8
COUNTRY: Italy / 意大利

Evnet — Here but Everywhere / Evnet —天涯共此时

DESIGNER: Tommaso Arcangioli, Gianfranco Bombaci, Lorenzo Castagnoli, Angelo Alessio Grasso

Environment   "Here but everywhere" seems to be the imperative of a technological revolution with an economic, politic and social weight without precedent. Information is not only power, but a real necessity to be part of a global society. Architecture can draw back from this revolution, not only using these new tools to comunicate itself, but to partecipate to this evolution and to make the ethereal informatic flow part of his concepts.

Contents   Evnet is an architectonical experiment to realize an event based on the network logics. The body of each partecipant is considered through his senses as an information receiver. This informations come from outside through mitters of sounds, images, scents, temperature. An informatic mix of sounds, images, scent and climatic conditions creates the artificial landscapes of Evnet. It is a "virtually" real experience.

System   Evnet is a silent party. Sounds come from different places, through radio, web radio, samplers. They are mixed by a master DJ and, sent on different frequencies, they are received by radio-headphones: partecipants can choose what they want to listen or without headphones they can listen just silence. Images come also from different places through a system of six cameras. They are mixed by a master VJ together with other images from web-cameras, projected on the walls as digital windows. The air conditioning system permits the regulation of temperature, related or not to the images (desert-hot/cold). Through areators scents and flavours are immitted in the hall stimulating also the sense of smell as receiver.

Result   Evnet is a space. It is the rappresentation of what we live today through an event: the creation of a landscape with no identity, in which space and time loose their meanings gaining new value.

环境　"天涯共此时"似乎是个具有经济、政治和社会价值却没有先例的技术革命的规则。信息不只是权力，更是成为世界公民的必备条件。除了运用这些新的工具来沟通外，建筑参与这个革命并以精致的信息潮流作为其概念的一部分。

要旨　"Evnet"是为实现一个基于网络逻辑的活动的建筑学实验。参与者的身体被认为是信息的接收器。信息来自于外界的声音、影像、嗅觉与温度。由声音、影像、嗅觉与温度组成的信息创造事件的人造景观。它是"虚拟的"真实体验。

系统　"Evnet"是无声的舞会。从不同地方来的声音，从广播、网络电台和取样器而来，由DJ掌控的混音，播放到不同频道，由广播耳机接收：参与者可选择想听的或不想甚至不听的。影像也通过六个摄影机的系统从各处传来，由掌控的VJ混和所拍摄的影像与网络摄影机所摄影像混和在一起，投射到墙面上作为数码橱窗。空调系统保持空间恒温，可依需求与影像产生关联(沙漠—热与冷)。通过喷雾器，把气味注入大厅刺激嗅觉。

成果　"Evnet"是空间。它通过一个侧面代表我们今日的生活：一个没有特定个性的景观的创造，在空间与时间失去意义的同时获得新的价值。　[翻译 蔡咏岚]

AFFILIATION: Nicole_fvr
COUNTRY: Italy / 意大利

Evnet_event: it is a silent party. People recieve by radio-headphones all the audio informations, they dance watching the web video inputs, feeling the weather conditions changes and smelling the scent jokey mix / Evnet- 活动: 无声的舞会。人们通过收音耳机接收所有的听觉信息，它们观看电视并且舞动，感受气候变化和闻着气味

Monument to the Third Millennium / 第三千禧年纪念碑

DESIGNER: William M. Marion, Herbert I. Burns, Mili Mulic

Monument to The Third Millennium   The competition was commissioned by the Governor of Puerto Rico who sought designs for an internationally recognizable monument dedicated to the third millennium. The site for this project is the Third Millennium Park, which is adjacent to the Atlantic Ocean in San Juan, Puerto Rico. The monument is designed to be highly visible and easily recognizable to persons arriving by land, by sea and by air. Furthermore, the monument was designed to incorporate a variety of digital and multi-media applications in a dramatically unique manner among the three architectural components as follows:

Digital Dirigible   The digital dirigible is comprised of a lightweight metal geodesic frame with a teflon coated mylar inner sphere. The dirigibles membrane also doubles as a digital projection screen for multi media presentations. The geodesic frame supports digital cameras available to an international audience via the internet and it supports an array of lasers for digitally monitored light shows. This sphere may be tethered; remote controlled, or piloted and is easily docked in the event of severe weather. Also, a limited number of passengers can be transported during special events.

Docking Amphitheater   The amphitheater is a masonry structure designed to compliment the San Geronimo fortress in scale and material. It also functions as the boarding location for passengers. The design includes forms and hieroglyphics that signify the Taino's relationship with the cosmos. Above all, the amphitheater hosts multi-media events or other open air activates.

International Digital Park   The digital park is a community of fully accessible paths and kiosks where large groups of visitors may assemble for special events. Visitors may view and interact with real-time digital images transmitted from monuments around the world that includes the pyramids at Giza, the Eiffel Tower, the St. Louis Arch, the Sydney Opera House, etc.

第三千禧年纪念碑   本次竞赛是由准备建造一个国际性第三千禧年纪念碑的波多黎各的领导人发起的。基地是位于圣胡安市邻太平洋的第三千禧年公园。纪念碑设计必须使不论从陆地、水上或空中来的人都可轻易辨认出它来。此外纪念碑的设计在三个建筑当中以戏剧化的独特性组合了各种数码与多媒体。

数码飞船   数码飞船由一个轻金属骨架和铁氟纶涂层的聚酯薄膜内球面构成。飞船的膜也作为数码多媒体展示的投影屏幕。在骨架上有通过网际网络发送给国际观众的多部数码相机，提供可由数码控制的灯光表演的极光阵列。飞船的膜可以栓住、遥控、驾驶，且当天气恶劣时可轻松停泊。有特别活动时可运载一定数量的乘客。

露天剧场   船坞露天剧场被设计成为赞扬圣·杰罗尼模堡垒的石造建筑。它也是旅客下船的地点。设计包括了一些象征着泰诺人（西印度群岛已绝迹的民族）宇宙观的形状与象形文字。最重要的，在露天剧场将上演数码多媒体活动或其他户外表演。

国际数码公园   数码公园内有全然可及的道路与凉亭，大量的访客可以在此集会活动。他们可以观看并且与从金字塔、艾菲尔铁塔、圣路易拱门或悉尼歌剧院等地传来的即时数码影像进行互动。[翻译 蔡咏岚]

←This view shows the interior of the docking amphitheater while the digital dirigible is in the docked position. In addition to its function as the docking station, the amphitheaters audience can is view multimedia presentations on the surface of the dirigible / 数码飞船停泊时的露天剧场船坞室内观景。露天剧场船坞除船坞的功能外，剧场的观众可以观看飞船膜面上的多媒体展示

AFFILIATION: MBM Studio
COUNTRY: U.S.A. / 美国

Real Scene — Virtual Space / 现实场景—虚拟空间

DESIGNER: Bert Chang / 张柏杨

Begin   We choose Chunghsiao E. Rd to be site. To watch, record, and imagine ... Fantastic, disappointed, expectant, and strange feeling from the city. Just want to give the city some energy for changing ...

Find   I become a stranger ,when I walked on the road. I use a camera to record something in the road, and simulate some scenes of my feeling by computer. Maybe our eyes can't ecognize these scenes, because these are in my imagination.

Imagine   "A.D.2612, in a spaceship, many people want to find another planet, because the earth can't suit human to live. The only survivors take this ship to galaxy..." This is a scene in my imagination. Why do I have the imagination? Maybe science movies give me many scenes of future. But science movies isn't real future. They are movie director's imagination. When people see similar scenes in daily life, they can compare movies and real life. After see many science movies, I want to make a scene which is belong my imagination. But future may not be this...

Final   The new skin connects everywhere in the area. Towers and bridges offer people to move. Information and image are everywhere on the skin. These are my way to change the city in virtual space...

开端   我们选择了中国台北忠孝东路作为基地。观察、记录、想像它是惊人的、是令人失望的、令人期盼的或来说是疏离的。只是想提供城市转变的能量。

寻觅   当我行走在街上时，成了一个陌生人。我用相机记录街道，并且运用电脑模拟一些感受到的场景。也许你们的眼睛无法辨识，因为它们存于我的想像之中。

想像   "公元2612年，在一艘太空船上，许多人因为地球不再适合人类居住而想寻找另一个星球。残存者于是驾着太空船在宇宙中寻找。"这是我想像中的一个场景。为什么会有这样的想法？或许是科幻电影给我的对未来的想像吧，它们是导演的想像。当人们在平常生活中看到类似的场景，他们将现实与电影相比较。看了很多科幻电影之后，我想创造出属于自己想像的场景。但未来不见得如我想像一样 。

最终   新的表面连接了区域中每个地方。高塔与桥梁供人们移动。信息与影像存在于表层每处。这是我改变都市虚拟空间的方法。[翻译 蔡咏岚]

←A virtual space in my imagine / 我想像中的虚拟空间

AFFILIATION: Commerce Department of Architecture Engineering, Chung Kuo Institute of Technology/ 中国台湾技术学院建筑工程系

Schizophrenic Dichotomization - Figure / Field /
精神分裂的对分—形体与场域

DESIGNER: Hung-ming Cheng / 郑鸿铭

Binary relations in architecture culture, as in the relations between image and form, trace a tradition leading to schizophrenic dichotomization. Through this idea might translate into an operational approach to the making of architecture might be much simpler than the paintings from which these ideas are derived. They can be seen to directly effect the way that the existing field (context) can be fully engaged. By seeing the field through the eyes of Analytical Cubism, by breaking it down to its elemental constituent components, the fullness becomes apparent. Interventions into that field have the ability to alter or reinforce the given relationships among the constituent components and the new intervening elements. Thereby, allowing new relationships between form, which was previously seen as an autonomous figure and space, now seen as a fully engaged figural element constituting field. In doing so. Simultaneous relationships and ambiguous readings of figure and field become possible.

By working with the voids as solids, it was possible to manipulate the form to create a completely scaleless and siteless, shallowless and deepless. The process of making not only promoted the skin and content as a spatial construct for future use in the studio, but also inspired a methodology of critical inquiry: manipulation of the given elements, union of these transformed elements to form a hybrid, and, finally, the intersection of this hybrid with a given context.

建筑文化中的二元关系，就如同影像与形体的关系，描绘出导向精神分裂对分的传统。通过此想法也许会转化为将比绘画更易于创作的建筑设计的操作。它们直接影响了现存涵构的组织。将之以立体分析法检视，分析拆解为组成构件，它的成熟度是可见的。介入涵构中即可改变的或强化现有的组成构件与新元素之间的关系。所以，以往被视为独立的形体与空间，现在被视为完全相关的形体元素而组成涵构，允许形体与形体之间产生新的关系。如此一来对同步发生的关系和对形体与场域模糊进行诠释成为可能。

借助于把虚空间当做实体的操作，才可创造出无尺度、无基地、无深度的形体。创造的过程不只将表层和内涵提升为未来工作室建构的空间元素，并且启发了这些变形的元素并使之形成一个复合体，并且最终形成此复合体与现存涵构的融合体。[翻译 蔡咏岚]

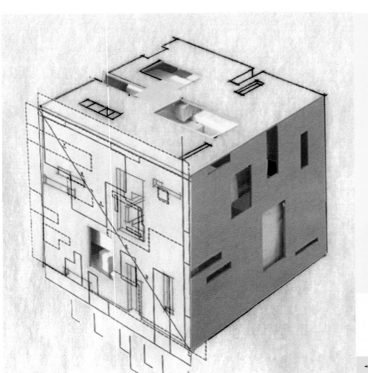

←Concept Model / 概念模型

AFFILIATION: The Chung-Kuo Institute of Technology / 中国台湾技术学院建筑工程系

Sand-City / 沙城

DESIGNER: Domenico Cannistraci, Pietro Chiodi, Matteo Costanzo, Valerio Franzone

An infrastructure that invades the spaces of the city, re-imagining urban landscape. This new urban soil contains life, pauses, information, nature, different kinds of events. A topographic surface is superimposed to the print of the city, now just a simple container of volumes.

Sand-City is a continuous city that merges itself with nature. It is in a state of mutation, showing the traces of the atmospheric agents, of everyday life, of the city itself. The citizen of the third millennium uses the urban space as a place for inter-relations, relaxing, doing sports, resting, relegating moments of activity and of working in the domestic environment.The city is not anymore the place of movements, of changing place, but the lounge where to consume life. To get back to the urban space by walking, by climbing it, by running, by slipping, by discovering it, by recognising it, by using it, by taking possession of it...

一个侵入城市空间的公共设施重新建构了城市景观。这新的城市土壤含有生命、断层、信息、各种活动与自然。在城市涵构上放置一个地形的表层，而它现在是简单的容器。

沙城是个不断与自然融合的城市。它存在于一种突变状态中，呈现出日常生活与城市本身环境介质的踪迹。第三千禧年的市民利用城市空间作为交流、放松、运动、休息的场所，是放逐工作与活动的时刻。城市不只是活动场所、变换场所，也是生活的公共空间。可以走、爬、跑、滑、发现、认识、使用或拥有都市空间。[翻译 蔡咏岚]

Sand-City Moscow / 沙城，莫斯科

AFFILIATION: Sand Architecture
COUNTRY: Italy / 意大利

goWittgenstein / Wittgenstein, 加油

DESIGNER: Thomas Hayde

Time did overtake the Wittgensteinhouse. It was built by the austrian architect Ludwig Wittgenstein in cooperation with Paul Engelmann and rests upon its base in third district vienna since seventythree years, to live up again now as the essence of a new building. a process that was: -examination of the groundplan in consideration of an analysis of E. Flach; filling up the model's negative-space to a box; space experiencing with Virtual Reality Modeling Language models; dematerializing and stacking of an evolved module-: was followed by a reaction to the location's situation in the city and thoughts in this design-task's contents. Today the Wittgensteinhouse is situated in an inhomogeneous city-planning context. The one-family house is sourrounded by dense blocks and is adjacent to a assurance-company's house that is much higher than its sourroundings. This situation demands a reaction, which in this project should be conceived as a counteraction. Wittgenstein wrote in his 《*Philosophical Examinations*》 : "It is only buildings of air, that we destroy, and we get back to the basis of speech where it was built upon". The Wittgensteinhouse in this project is to be interpreted as the foundation of a skyscraper which is segmented by a module that evolved in a working process as explained above. This basis would function as a philosophical centre. Exhibitions, lectures, thoughts and discussions. The space of the original structure of the wittgensteinhouse has been emptied and thereby been opened for its new function. It supports the building above which contains offices and homes.

　　光阴确实是追上了Wittgenstein屋。73年前由奥地利建筑师 Ludwig Wittgenstein 和 Paul Engelmann合作建成的坐落于维也纳的第三区的房屋，现在将成为一个新建筑的基础。这是考察对 E.Flach 的分析后检查配置的过程，也是填充模型的负空间，虚拟模型和语言模型的实验，相关模具的堆叠与非物质化的过程，伴随着当地都市状况以及对此设计任务的思考。Wittgenstein 屋处于非同质的都市规划脉络中。这个单一家庭住宅被密集的街道包围，而且与一栋高于周围建筑物的保险公司办公大楼为邻。我们对于这样的局面需要做出反应，就这个案例来说，它应该被当做一个回应。Wittgenstein 在他的《哲学实验》中写到："我们被摧毁的仅是建造在空气中的建筑，而我们将回归到它赖以建造的语言基础。" Wittgenstein 屋在此案例中被当作一个摩天大楼的基础来诠释，此摩天大楼的分割牵涉上述的工作程序。Wittgenstein 屋的原有构造被腾出来做为哲学中心。有展示、演讲、思想研究与讨论活动。它服务了高层的办公室与住宅。

[翻译 蔡咏岚]

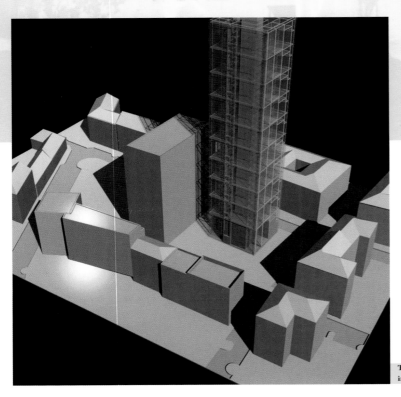

The houses structure was filled up to build a stackable module. This picture visualizes the "positive-negative-body" in its modular form / 房屋的构造被填充以建造一个可堆叠的模具。此影像将正 - 负 - 体的模矩形体视觉化

AFFILIATION: Student at Vienna University of Technology
COUNTRY: Austria / 奥地利

Constraint Based Space Planning — A Case Study /
在限制条件下的空间规划—案例研讨

DESIGNER: Ying-chun Hsu

The initial design process usually requires a long incubation time for architects to bring together form and function. In this exciting and painful process, architects use their professional training to adjust the relationship between spaces, and include the requirements from the clients.

The computer has the potential to be the key and the most powerful tool in this long process. In the software market, the software packages for architecture now are mostly for computing values or drafting. They are not very helpful in the design process. After analyzing design methods and comparing the potential of computer languages, I found that some computer languages have the possibility to virtualize some levels of the design process. One area of design that I investigated is space planning. Based on collected space relationship data a space diagram can be developed. In an ideal process, this result could shorten the planning time for architects and let them concentrate on design and details. AutoLISP within AutoCAD 2000 was selected as the programming language. To determine the logic for the program, the design process was compared with the computer process and it was determined that the program should begin with a foundation consisting of a floor plan matrix. After the matrix is formed, the program was further developed so to able to recognize the shape and size of the site, its conditions, and the location of spaces. According to this data, the best possible positions for the spaces can were attempted to be found.

在设计过程的初期，建筑师通常会花长时间结合机能与形体。在令人兴奋却烦恼的过程中，建筑师以其专业技术调整空间之间的关系，当然还要考虑委托人的要求。

电脑有可能成为这漫长过程中最有利的关键工具。软件市场中大都是计算数据或绘图的应用程序。这些对设计帮不上忙。在分析了设计方法与比较了各种程序语言的潜力后，我发现一些可能将某程度的设计过程虚拟化的语言。其中一项是空间规划。基于搜集到的空间关系数据，可绘出一个空间图表。在理想的过程中，这种结果可以缩短建筑师计划的时间，而让他专注于设计与细部上。AutoCAD 2000 中的 AutoLISP 被选作为程序语言。为决定程序的逻辑，设计程序被拿来与电脑的计算程序做比较，由此决定了程式应该由一个包含楼层平面的矩阵开始。构成矩阵后，进一步发展程式到能够辨识基地的形状与大小、形况、空间的位置。最后根据这些数据，找到空间最合适的位置。[翻译 蔡咏岚]

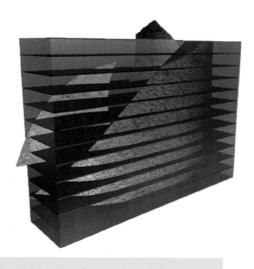
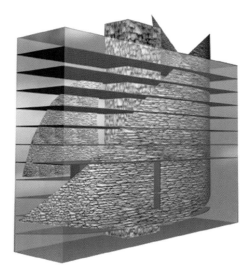

AFFILIATION: Ph. D Candidate, College of Architecture, Illinois Institute of Technology, Chicago U.S.A.
COUNTRY: U.S.A. / 美国

Vision for the City in the Third Millenium / 对于第三个千禧年城市的憧憬

DESIGNER: Monika Koeck, Dipl.-Ing.(FH)

The "city" will no longer exist as a well-composed conglomerate of cultures and professions. It will be a highly specialized place where everything is focussed on the purpose of the city, its industrial or technological aspects. Just like for example Silicon Valley. There will be cities that will no pay attention to our natural resources and will eventually destroy themselves. To offer a better life for its inhabitants, there will be cities created that pretend to be paradise, but for what price? For suffering at our work place, the city of technology, being isolated and lonely? For living in an artificial and not grown place with bad urban planning? The visions for the city in the future are negative and disastrous. Hopefully the city will create its own vision, and hopefully this vision will be a composition incorporating all aspects of urban living, the positive and the negative in a well-balanced society.

城市将不再是圆满建构的文化与专业的集合体。它将是高度专门化的地方，每件事事物都着重于城市的用途以及工业或科技的方向。以硅谷为例，将会有许多不考虑自然环境的城市出现，但它们终将自我毁灭。为给予居民更好的生活，将会有所谓的城市乐园被建造出来，但代价是什么？为了向我们提供一个隔离的、孤独的，不舒适的科技城作为工作场所？为了居住在非自然发展而形成的规划粗劣的人工环境中？对于未来城市的憧憬是负面且毁灭性的。但希望城市将为自己制造憧憬，更期望这憧憬将是体现都市生活各种观点的构图，正面的东西和负面的东西将会在社会中获得平衡。[翻译 蔡咏岚]

↑ Dead City / 鬼城   ↑ Industrialized City / 工业化城市   ↑ Technological City / 科技的城市   ↑ Interlaced City / 交织的城市

AFFILIATION: Visiting Assistant Professor, College of Architecture, University of Oklahoma
COUNTRY: Germany / 德国

Kinetic Collage, Digital Architecture — Smallest Space for a Meditation

体育学院—数码建筑冥想的最小空间

DESIGNER: Hirotaka Koizumi

I only want to see the reverse side of the world that we believe "real."

Which did Lewis Caroll image in the "Through The Looking-glass and What Alice Found There" and Enric Miralles in the project of "Small Wooden House."

我只是想要看到我们所确信为"真实"的世界的反面。

是Lewis Caroll在"透过镜子和爱丽丝在那里所找到的"以及 Enric Miralles在小木屋的案例中想像的。[翻译 蔡咏岚]

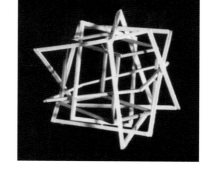

→**Perspectives: 3 Rectangles give the 2.0×2.0×2.15m cube a scene full of variety, and would become rich the interior image by means of 3 electromagnetic screens** ／透视图：三个矩形赋予 2.0m × 2.0m × 2.15m 的多样的场景，室内因三个电子屏幕而丰富起来

AFFILIATION: Hirotaka Koizumi Architects Associates
COUNTRY: Japan / 日本

Establishing Experimental Media Ground / 确立实验性媒体的立场

DESIGNER: Junya Sakai, Ferry Gunawan

We begin our design by setting up conditions in which subway platforms would be designed and controlled by the MTA, sponsoring corporations, and individual donors. Central to this setup, is the establishment of a non-profit foundation that organizes the competition for the design of the platform. This setup is conditioned to create opportunities to convert portions of various public spaces in the city, in this case, a subway platform, into an experimental ground that requires critical communications between, government, corporate, and non-profit foundations. The main task for the foundation is to intervene in the design of the platform to further beatify its state and provide continual funding for the competition. Throughout the design process, digital architecture is apparent, from delivery of designs, to design of programs, to production.

The delivery of the design begins with holding a competition to determine the design of the programs that will be utilized on the platform such as digital displays, Internet connections, and light fixtures. These devices will be standardized and mass-produced to reduce the cost of production. Shop drawings of these devices will be sent electronically to all applicants prior to the beginning of a second competition that will determine the actual design of prototypical forms that will integrate these devices. The application of the design, primarily in the design of programs, such as a NYC neighborhood directory, commercial advertisements, and an emergency care response system, will have a real-time digital connection to the information network.

The production and process of design must take advantage of 3D modeling software and be able to feed into automated machines such as the CNC milling machine that can read computer files. This is intended to experiment with the potential of computerized design as well as to evaluate the efficacy of computerized production. The maintenance of the design as well as allocation of collected funding will be the responsibility of the MTA. Ideally, every 5-10 years there will be an evaluation on the effectiveness of this overall process based on the cost of production and the response of people to the design. The platform can become an experimental ground for electronic device manufacturers as well as for platform designers, fostering improvements and advancements.

我们的设计从设定地铁月台的环境开始，该地铁月台将由 MTA、赞助人、独立捐献者设计并控制。在这种布局的中心是设立一个安排地铁月台竞赛的非营利基金会。这样的布局是为了把都市中的各种公共空间，（在此指的是地铁月台）转化成为需要政府、法人和基金会间紧密沟通的实验性场所。基金会的任务是介入地铁月台的设计，进一步督导它的状况并持续资助竞赛。在设计的过程中，从设计的传达、计划的拟定到制作，数码建筑随处可见。

设计的传达始于举办一个竞赛来决定计划的拟定，例如利用月台上的数码显示器、网络连线和灯光设备。这些装置被标准化且大量生产以降低制造费。大样图在举办第二个竞图前就会以电子方式寄送给参与者，这是为了结合各设备原型形体的实际设计。设计的运用，首要是计划的拟定，例如纽约市的邻里通讯、商业广告和紧急看护反应系统，会有即时数码连线到信息网络上。

设计的生产和进程必须利用 3D 模型的软件，而且要能输入自动化的机器中，以实验电脑化设计的潜力和评估电脑化制造的效率。设计成果的维护和募得捐款的分派将由 MTA 的责任。在理想状况下，每五到十年会有根据成本与人们反映所作出的整体效率的评估。地铁月台将成为电子设备制造商和设计者促成改进与升级的实验场所。[翻译 蔡咏岚]

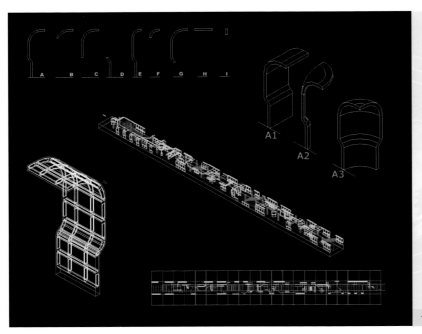

←Prototypes and their final layout / 原型的最终配置

AFFILIATION: Walker Group/CNI
COUNTRY: U.S.A. + Indonisia / 美国 + 印尼

Angel 2035 / 天使 2035

DESIGNER: Sigbert Schindler (German) / (德裔)

The house of the future is the house of the present studying the past like a laboratory (Exploratorium) to go to the future. The research of new ways of life guides to new ways of construction whether its on the earth or on the Mars. The room in the information age is determined essentially through 3 conquests: the human body, the space, and the virtual room. The house of the future then is an open center for the physical - mental and cultural potential of the human being and his creativity. The future dwelling of tomorrow will get his essential quality through his technical and especially "social" installations. It will be produced in large numbers of pieces in modular designs.

Underneath the main axis of this center you find a museum which is consisted of a linear tube being at the same time a projection screen and a roadway for the artist working in speed. Interactive together with the "Hypermovie-Angel-Conference-Hall" and the "Moving-Container-Ateliers" it forms a main quality of "Angel 2035". The "Living-Pneu-Wings" connected with the main axis let their inhabitants find their individual pleasure. The softly inclined "Angeltower" is open for the most different collective events. The "Office-Question-Buildings" contains Hologramscreens which allow you to create at the same time a nice clima ("screen-air-conditioning"). In the "Hypermensa" everyone can refresh himself with a nice bath and a meal.

A further particularity of "Angel 2035" is the social and cultural Trinity which is formed through the "Moving-Container-Ateliers", the "Hypermovie-Angel-Conference-Hall" and the "Exploratorium". The combination of economy, science and culture! I wanted to create a center within empty spaces where the "Dancer" (human being) the "Prophet" can leave his prints in the present for the future. A center like the life moving towards all directions.

未来之屋是迈向未来，存在于现在的实验室。新生活方式的研究引领新的建造方法，不论是在地球上或火星上。信息时代的居室建造应基于三个挑战:人体、宇宙和虚拟居室。未来之屋将是释放人类及其创造力的肉体精神和文化潜能的中心。未来的住所将以其技术的和社会的设施为特色。它将以数目庞大的模块化方式生产。

位于这个中心主轴线的底端的是一个被视为既是投射屏幕和快速工作艺术家的通路也是线型馆的博物馆。它与"动态电影天使会议厅"和"移动容器工作室"两者互相影响，共同形成"天使2035"的主要特质。与主轴线连接的"生活的精神侧厅"让居民找到各自的乐趣。微微倾斜的"天使塔"开放给各种公共活动。"办公室疑问建筑"让你操控产生好气候的全屏幕。而在"动态圣坛"里你可以沐浴和用餐。

"天使2035"的另一个特点是通过"移动容器工作室"、"动态电影天使会议厅"和"探索的集会厅"形成社会与文化的三位一体，是经济结构、科学与文化的结合体！我要在虚拟空间中创造的一个"舞者"和"预言家"，可留下供后世发展的印记。一个全方位的中心。

[翻译 蔡咏岚]

←Angel 2035 Center Overview ／ 天使2035中心概观

COUNTRY: Italy / 意大利

The Screen's Urban Reflection / 屏幕中的城市映影

DESIGNER: Dounas Theodoros, Kassios Apostolis, Tsopanoglou Giorgos

The site chosen for the competition is a city block located in the center of Thessaloniki, Greece. Its particular interest is the presence of significant remains of the Roman horse track (-7,00m below pavement level) combined with the fact that it is a generic example of an urban landscape. A small portion of the site remains unbuilt although the rest is taken up by multi-unit housing buildings. The unbuilt part of the city block remains void responsible of which is the existence of the remains of the roman horsetrack causing quarels and lack of interest in the local society.The empty site will be taken by a public center for digital media , by this way combining the wonderfull engineering roman age and the digital world .

Roughly the composition consists of a main volume of surfaces integrated with one another, beginning from a level of +7,00 meters above street level. An important role in the composition is played by "a movement on axis" consisting of stairs descending to the level of the Roman Remains; entering the building and leading through to the main "yard" of the block, where an open air cinema (that can be transformed into an interactive display with the click of a button) is located. Projection via electronic media is a fundamental element of the building. Presence of computer screens in virtually every part of the complex and a stormy network transforms the composition from part of the built environment into a platform of interactivity, digital media and information. Every imaginable way of communication and social activity becomes possibility within the premises of the block because even the pre-existing buildings are connected to the network of the new building.

为竞赛所选择的基地位于希腊Thessaloniki市中心的一个街区。它的特点是基地上独特的古罗马马车道路遗迹(在路面七米之下),还有都市景观的典型。基地的一小部分仍未开发,其余则被复合单元的住宅占据。遗迹通过留下的空地引起当地社会的争执。未被占用的基地将是一个公共数码媒体中心,以此把古罗马时代精彩的工程与现代数码世界结合起来。

概要地说,此作品由一个有数个面的主要量体构成,从高于街道七米处起始。"轴线的运动"扮演重要的角色,它包含了可以下到遗迹位置的阶梯;进入建筑内,穿过本街区的庭院,在那儿有一个(通过一个按键就可变成一个相互作用的表演)户外剧场。利用电子为媒介的影像投影是建筑上一个重要的元素。此复合方案内随处可见的电脑屏幕以及猛烈包围的网络把它从已被建造好的环境的一部分转化为活动、数码媒体和信息相互作用的平台。因为连先前存在的建筑都会连接上新建筑的网络,所以在街区范围内将发生各种可能的沟通和交流活动。[翻译 蔡咏岚]

← Section of the complex showing the main building to the right, open air cinema to the lower left and existing structure (built in the 50s-60s) for multiunit housing / 剖面图: 右方是主建筑物, 左下方是户外剧场和现存的建于20世纪50至60年代的复合单元住宅构造

AFFILIATION: Student (undergraduate), Aristoteleio University of Thessaloniki, Polytechnic School, Department of Architecture
COUNTRY: Greece/ 希腊

Museum/ 博物馆

DESIGNER: Maurizio Mazzoli

The Museum is composed of an exhibition hall (a long parallelepiped with two side towers made of wood and glass sheets), a concert room, a conference room, an open-air exposition area and services, it stands as a close interchange between water and light, an interplay between full and empty areas and see-through effects. It is a steel structure, made of wood and structural glass erected on a water sheet. The main front looks like a moving fluid, the light draws figures and geometric shapes. There is an explicit reference to industrial architecture, filtered and reinterpreted, fluctuating between immateriality and presence. Every space is independent but it is also an interactive part of the whole structure, which is protected by steel brise-soleil. The water surface pierces and integrates and it becomes an active element. The entry to the gallery is characterised by a wide web of indirect light with a water sheet, which can be easily seen by the impluvium-place. The sun light becomes a scenario: in the concert room the spectators-performers will see the chronological sequence of the moving rays stressing at their backs, in a sort of shiny clock-seabed. The light is a structural element.

　　博物馆由展示厅（平行长廊和两侧木材玻璃制的塔）、演奏厅、会议厅、户外解说空间和服务空间组成，它因为与光紧密的交互作用、实空间与虚空间、穿透空间之间的相互作用而存在。它直立在一张水做的薄面上，由钢、木材与结构玻璃所构筑。正面看起来像流动的液体，光线在其上描绘了各式形体与几何形状。对于工业建筑有明确的了解，经过过滤与诠释后，在非物质与实体间波动。每个空间都是独立的，却都是受钢材保护的整体的一部份。穿透水面结合后，成为一个活跃的元素。在冷色调的场所里可轻易看见以一个间接光线与水面交织成的宽大的网作为特征的展示厅入口。阳光成为一出戏剧的剧本：在演艺厅里的观众与表演者将看见移动的光束依时间顺序在他们背后一连串地表演着，就像时钟光亮的底部般。在这里光是一个结构性元素。[翻译 蔡咏岚]

AFFILIATION: Architech _Italian Design Group
COUNTRY: Italy/ 意大利

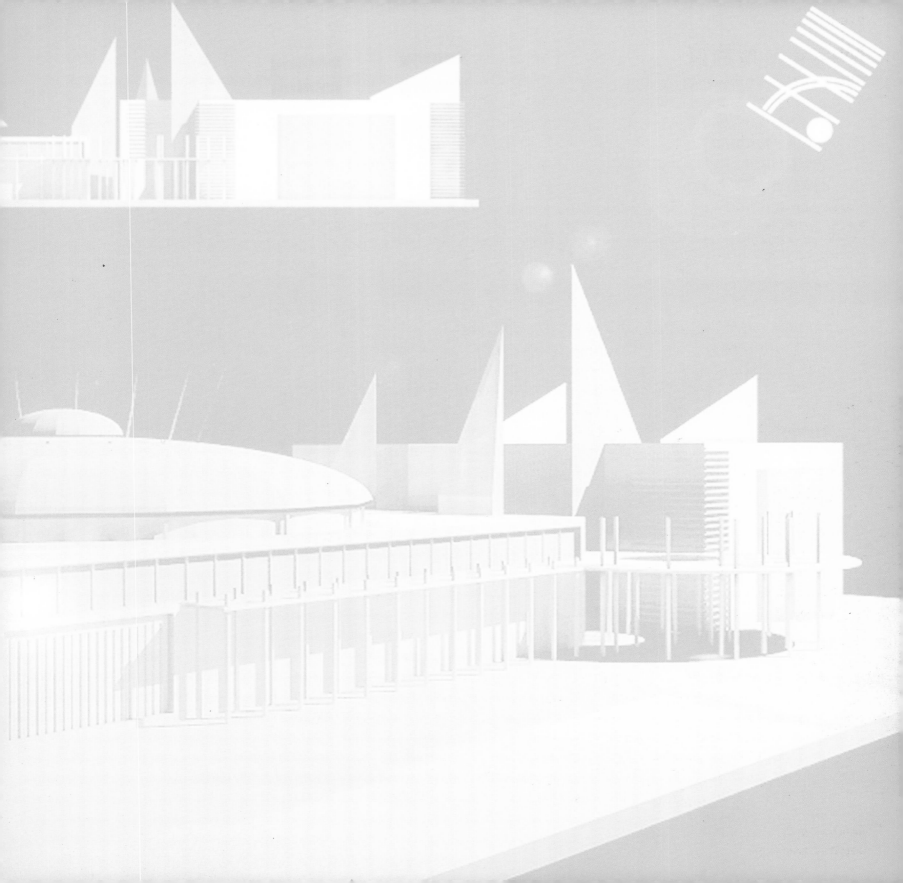

图书在版编目（CIP）数据

数码建筑／刘育东编 . —大连：大连理工大学出版社，2002.7
ISBN 7-5611-2005-2

Ⅰ.数… Ⅱ.刘… Ⅲ.建筑设计－作品集－远东－现代 Ⅳ.TU206

中国版本图书馆CIP数据核字（2002）第001589号

大连理工大学出版社出版发行

大连市凌水河　　　　邮政编码：116024
电　　话：0411-4708842　　　传　　真：0411-4701466
E-mail: dutp@mail. dlptt. ln. cn
URL: http://www. dutp. com. cn
印　　刷：利丰雅高印刷（深圳）有限公司
开　　本：787 毫米×1092 毫米　1/12
印　　张：19.5
印　　数：1—3000 册
2002 年 7 月第 1 版
2002 年 7 月第 1 次印刷
出　版　人：王海山
责任编辑：柳战辉　苍　峰　刘　蓉
责任校对：田茂林
封面设计：王复冈
定　　价：180.00 元